The Wardian Case

The Wardian Case

How a Simple Box Moved Plants and Changed the World

Luke Keogh

The University of Chicago Press

Kew Publishing
Royal Botanic Gardens, Kew

The University of Chicago Press, Chicago 60637
© 2020 by The University of Chicago
All rights reserved. No part of this book may be used or reproduced in any manner
whatsoever without written permission, except in the case of brief quotations in critical
articles and reviews. For more information, contact the University of Chicago Press,
1427 E. 60th St., Chicago, IL 60637.
Published in 2020 by the University of Chicago Press
and
Published in 2020 by the Royal Botanic Gardens, Kew, Richmond, Surrey, TW9 3AB,
UK, www.kew.org
Printed in the United States of America

29 28 27 26 25 24 23 22   2 3 4 5

University of Chicago Press ISBN-13: 978-0-226-71361-8 (cloth)
University of Chicago Press ISBN-13: 978-0-226-71375-5 (e-book)
University of Chicago Press DOI: https://doi.org/10.7208/chicago/9780226713755
.001.0001
Kew Publishing ISBN-13: 978-1-84246-719-0

Library of Congress Cataloging-in-Publication Data

Names: Keogh, Luke, author.
Title: The Wardian case : how a simple box moved plants and changed the world /
    Luke Keogh.
Description: Chicago ; Royal Botanic Gardens, Kew : University of Chicago Press ;
    Kew Publishing, 2020. | Includes bibliographical references and index.
Identifiers: LCCN 2020006507 | ISBN 9780226713618 (cloth) |
    ISBN 9780226713755 (ebook)
Subjects: LCSH: Ward, N. B. (Nathaniel Bagshaw), 1791–1868. | Wardian cases—
    History. | Plants—Transportation—History—19th century. | Wardian cases—
    Environmental aspects. | Globalization—History—19th century.
Classification: LCC SB417 .K46 2020 | DDC 635.9/8—dc23
LC record available at https://lccn.loc.gov/2020006507

A catalog record for this book is available from the British Library.

♾ This paper meets the requirements of ANSI/NISO Z39.48-1992 (Permanence of
Paper).

For Apple

Contents

....

Color illustrations follow page 160.

THE WORLD: SHOWING SHIPPING ROUTES AROUND 1850, AND SOME OF THE MANY PLACES THE WARDIAN CASE TRAVELED.

Introduction

....

The first journey of a Wardian case was an experiment. In 1829 the surgeon and amateur naturalist Nathaniel Bagshaw Ward accidentally discovered that plants enclosed in airtight glass cases can survive for long periods without watering. After four years of growing plants under glass in his London home, Ward created glazed traveling cases that he hoped could be used to transport plants around the world. The small, sturdy cases were made of wood and glass and looked a lot like portable greenhouses. In 1833 he tested his invention by transporting two cases filled with a selection of ferns, mosses, and grasses from London to Sydney, the longest journey one could take at the time.

On 23 November 1833 Ward received a letter from Charles Mallard, the ship captain responsible for the two cases: "your experiment for the preservation of plants alive . . . has fully succeeded."[1] The next challenge was the return journey. In February 1834 the cases were replanted with specimens from Australia. Eight months later, when Ward and his friend George Loddiges, a well-known nursery owner, went aboard the ship in London, they inspected the healthy fronds of a delicate coral fern (*Gleichenia microphylla*), an Australian plant never before seen in Britain. And along the way a few black wattle seedlings sprouted in the soil. The experiment was a success.

The Wardian case, as it became known, revolutionized the movement of plants around the globe. After the first experiment, thousands of cases were used over the next century to move plants. It was a "simple but beautiful" invention, as one sea captain described it. The first to use the cases on a large scale were the commercial nursery firms, who saw the value of the technology and quickly began to send the cases out. Then the world's leading botanical institutions, such as the Royal Botanic Gardens, Kew, and London's Royal Horticultural Society adopted the case to their needs.

Explorers, missionaries, plant hunters, and government officials used the Wardian case, as did all nations with an interest in plants. One robust horticulturalist working on the United States Exploring Expedition (1838–42) collected plants and packed them in Wardian cases; he landed them home after four years of travel, and they laid the foundations for the United States Botanic Garden in Washington, DC. Over the next half century important agricultural plants such as bananas, cocoa, rubber, tea, and many more were successfully moved in cases and went on to have major commercial impacts. The effects of these impacts were significant and widespread.

This book tells the story of the Wardian case (fig. 1) from its invention in 1829 to its final significant journeys in the 1920s. Specifically, the book is about the worldwide movement of live plants. The Wardian case played an important role in moving all variety of plants and, for good or ill, helped to transform the world we live in today. Many scientists, historians, and garden writers acknowledge the importance of the Wardian case in moving plants around the world, but the full story that charts its long and useful history has never been told.

The Wardian Case tells a century-long story of people moving plants. Subtly it shows how our perceptions and relationship to nature changed with our ability to move plants effectively. As well as charting a much fuller story of the Wardian case, this book centers on two important themes: possibilities and environments. As a technology, the Wardian case provided the possibility to move plants; as an enclosed case it moved more than plants—it moved environments.

FIGURE 1 Wardian cases preparing to leave the Royal Botanic Gardens, Kew, ca. 1940.
© The Board of Trustees of the Royal Botanic Gardens, Kew.

Plants, Possibilities, and Practice

Before the invention of the Wardian case, transporting live plants was difficult and costly. Of course seeds could be sent but the seeds of many species will die if they dry out or become moldy if kept damp, especially those that are oil rich or from the tropics. Furthermore, sometimes the season was not right for collecting seeds. Therefore, sending live plants was the most viable option, but before the invention of the Wardian case this was difficult. In 1819 John Livingstone, a keen botanist and surgeon posted to Macao for the East India Company, wrote about the challenge of sending live plants from China to London.[2] Livingstone estimated that only one in a thousand plants survived the journey. He concluded that it "becomes a matter of importance to attempt some more certain method gratifying the English horticulturalist and botanist, with the plants of China." As shipping increased and the world became more connected through exploration and trade, there was a need for a better method to move plants. The Wardian case filled this niche.

The Wardian case did not always go by the name it has now. Variously they were called closely glazed cases, glazed cases, and glazed boxes. Not until the 1840s was the name "Ward's case" or "Wardian case" used. The nomenclature might seem to be a small detail, but it bears on the story of the case. Nathaniel Bagshaw Ward was born on 13 August 1791 in Whitechapel, London. His father, Stephen Smith Ward, was a medical doctor who ran a practice at Wellclose Square. The family home was not far from the London docks, and young Ward spent much time there collecting curios such as matchstick boxes. From a young age he was an avid amateur naturalist. In the late eighteenth century few but the upper classes could afford to spend all their time on plants. Ward joined the family profession: he trained as a doctor and later assumed his father's practice in London's East End. On 4 September 1817 Ward married Charlotte Witte, and the two assumed residence at Wellclose Square. They had as many as nine children. From the few sources that exist, it is hard to know just what Ward's family thought about all the plant experiments, but certainly both Charlotte and the children took part in them and must have helped him maintain his many specimens.

While training as a doctor Ward spent time at the Chelsea Physic Garden—the garden of the Society of Apothecaries. Up until the late nineteenth century all medical doctors were required to have a thorough knowledge of plants, because it was from them that most remedies were derived. Doctors even had to pass a botany examination before they could complete their training; well into his sixties Ward oversaw newly minted doctors' botany examinations, which often fell on his birthday. He remained connected to the Society all his life and in later years even became Master of Apothecaries. Ward may have been a doctor by profession, but his passion was plants and gardening.

For an amateur he had a wide influence. His list of correspondents included the leading naturalists of the nineteenth century, including

Charles Darwin, Michael Faraday, Asa Gray, William and Joseph Hooker, and John Lindley. The success and quick uptake of Ward's invention for moving plants cannot be separated from his networks. For example, his close relationship with the nurseryman George Loddiges, who supplied the plants for the first experiment in 1833, was an important element in its success. Ward made three important contributions: he hit upon the idea of the enclosed system, he experimented with the case by sending it to Australia, and he promoted it widely through his networks. For this he deserves recognition.

Although the name suggests otherwise, the Wardian case was about much more than Nathaniel Ward. As historians of technology have repeatedly shown, inventions are hardly ever born in an "aha" moment: there is always a longer culture of improvement at play.[3] Both before and after Ward there were significant turning points in using plant boxes. Ward's moment of discovery, although good fodder for folklore, cannot be understood without a longer look back into the eighteenth century, when moving plants in boxes was a common practice. Both British and French botanists pioneered plant boxes. Indeed, some of their suggestions, such as having an accompanying gardener and providing cross-ventilation, persisted well into the twentieth century.

After Ward's death the case named for him traveled more frequently and with greater impact. In this late phase the commercial nursery trade, late nineteenth-century imperialists, and even entomologists used the Wardian case and had significant impacts on the global environment. This panoramic view of the Wardian case is an important contribution of this book. It helps to show that technological innovation is always an incremental process and one that outlives the pioneering inventor.

Nathaniel Ward never patented his invention, and throughout its life the Wardian case came in many designs and styles. It is very important to distinguish between two types of Wardian cases: ornamental cases and traveling cases. This book is entirely focused on the latter. Both varieties of case bore the same name, but they were very different. Ornamental cases, precursors to what we know today as terrariums, were a common indoor feature of many Victorian-era homes, in which they were used to house delicate and beautiful plants. In the mid-nineteenth century ferns were highly prized and were a common variety to keep under glass. By the end of the century other species, such as orchids, were also greatly valued. And recently there has been a resurgence of this indoor botanical style, which dates back to the time of Ward in the early nineteenth century.[4]

While the ornamental case was designed with beauty and aesthetics in mind, the traveling Wardian case was sturdily built to withstand a long sea journey. It was made of strong timber; some of the first ones were built with teak from old East India Company ships. The sloping roof of the case held glass inserts. Over the top of the glass there were often wire gauze or timber battens to protect the glass while still allowing light to enter. Inside the case one could send plants in pots or in soil directly placed on the bottom. Either way, battens were fastened internally to keep the plants and soil secure.

In 1884, despite half a century's experience in using Wardian cases, sending one still demanded a lot of skill. Joseph Hooker, the director of Kew Gardens, wrote to his correspondent in New Zealand, "The fact is that filling Wards cases . . . is an *art* that requires both knowledge and experience. . . . What you want is, a thoroughly good practical gardener."[5] Hooker spent much of his career seeking professional recognition for botany, but the art of packing a Wardian case required the skills of a practical gardener.[6]

Of all the users, the nursery trade was not only the first adopter but also the key innovator of the Wardian case. The role of gardeners and horticulturalists as important actors in the global distribution of plants deserves greater recognition. The historian of science Phillip Pauly, in his *Fruits and Plains* (2008), showed that horticulture was no mere "ornamental subject."[7] To follow the Wardian case is also to see the wide impact of horticultural science.

Commercial nursery firms were quick to see the value of the Wardian case and to use it in their plant transfers. Exotic plants were big business in Europe and settler societies. Having access to a regular supply of interesting and unique foreign plants and distributing them effectively was an important part of maintaining a nursery business. Often forgotten in studies of major botanical institutions, such as Kew Gardens, is their intimate connection to the commercial nursery firms of London, Europe, and the rest of the world. During much of the nineteenth century there was a smooth and largely unrestricted flow of plants between institutions, nations, firms, and private citizens.

Nurseries, missionaries, ship captains, and merchants significantly impacted how and where plants moved.[8] The Wardian case shows how widespread civil society, as a group of users, influenced global plant movements outside, or on the edge, of the functions of the state, demonstrating the collaborative relationship between imperial science and science directed by amateurs. In the cases' first decade of use, it was nurseries, amateurs, and gardeners who first utilized it and proved its value. Indeed, French scientists only saw the worth of the case in 1836 after the nurseryman Loddiges's cases were sent to Paris. The hard work of moving plants was done not only by horticulturalists and botanists, but also by gardeners, colonized peoples, indentured workers, and slaves.

The Case of Earth

Following a box on its journeys allows a radical and unique way of seeing the history of environment to emerge. It not only provides a deeper understanding of how things move, but also hints at the interconnectedness of modern life and the global trade networks that shape this relationship. Mark Levinson employed such reasoning in his study of the modern shipping container and showed that it was no boring box. Similarly, the story of the Wardian case reveals new insights into a well-established line of inquiry. The Wardian case was not merely a conveyance; it was a "prime mover" that witnessed a major change in the way the global environment operated.[9]

The archival material relating to the Wardian case is fragmented and widely spread. Often botanists, nurseries and botanical institutions did not think to record details of the methods used for moving plants. Owing to the nature of the case and the varying source materials, I have chosen to illuminate significant users from around the world. By tracing evolving connections and interdependence, this book is in some ways a global history from the bottom up.[10] Throughout there is a large focus on Britain, because the case was invented there and because Kew Gardens was such an important botanical institution in the nineteenth century. But it is often forgotten that other nations were also important users of the case. I have tried to capture wider examples, including those of France, Germany, the United States, and Australia. In the late nineteenth and early twentieth centuries, all of them used the case more than Britain did and were far more innovative regarding its possibilities.

At its foundations *The Wardian Case* is a work of environmental history. It shows how dramatically our methods and attitudes toward moving plants changed over the course of a century. In the 1830s, when Ward was promoting his ideas, having an exotic plant sitting on your dining table was something to prize and marvel at. Over the next century the free movement of plants saw not only exotic curiosities arrive on foreign shores but also invasive species, diseases, and pathogens. Controlling such problems led to stricter quarantine and paved the path toward many of the practices of environmental management and biodiversity conservation that we have today.[11] Charting the fuller environmental history of the Wardian case reveals that its last major use was for biological control—to solve problems by moving insects that would feed upon out-of-control invasive plants. Such an ironic conclusion to Ward's wonderful invention should not be lost on anyone with a concern for long-term environmental change.[12]

Nature is always dynamic and mobile. Our histories should reflect this. Logistics is largely missing in much historical writing about the environment.[13] By focusing on a box that moved plants, we can get a global picture of plants on the move that collectively went on to have a fundamental impact on modern history. At its height in the nineteenth century there were thousands if not tens of thousands of these cases in operation, moving plants around the globe. Our choices of what we drink, eat, smell, and wear have all been transformed by the movement of plants. One object witnessed all of this: the Wardian case.

Out of the Attic

At the Botanic Garden Berlin there was a room in the attic full of disused gardening equipment. In the spring of 2010, while tidying it up, curators

FIGURE 2 The Wardian case on display at the Deutsches Museum, Munich, for
the exhibition *Welcome to the Anthropocene: The Earth in Our Hands* (2014–16).
The case is from the collection of the Botanic Garden and Botanical Museum Berlin.
Photo courtesy of Deutsches Museum, Munich.

found a Wardian case. The Berlin case was a rare find: it was one of the
very few known to exist in continental Europe and only one of fifteen
genuine cases known to exist globally. A few years after this chance
discovery, I traveled from Munich to Berlin to see the case while in my
role as curator working on the exhibition *Welcome to the Anthropocene:
The Earth in Our Hands* (Deutsches Museum, Munich, 2014–16; fig. 2).
I was trying to put together a section on mobility. I had already organized a feature on the shipping container, a sure symbol of human global
mobility patterns, but I wanted to take people further back to the roots
of these patterns. This was the first Wardian case that I had ever seen.

As I caught the train home from Berlin after seeing the case, I thought
about how my own life was impacted by the Wardian case. I regularly buy
bananas at the supermarket. Just the day before I had changed the tire on
my bicycle. I enjoy sipping Darjeeling tea but usually tend toward coffee.

When selecting a floral gift, I will buy chrysanthemums over orchids. The one impact I didn't so much enjoy was weeding my little garden. The Wardian case moved all of these plants and was instrumental in putting them in regions that led to their prominence.

After seeing the case in Berlin, I knew I had to put it in the Anthropocene exhibition. The Anthropocene concept suggests that the impact of humanity has become so great that we are now creating our own geological age.[14] The concept has inspired a range of people, from scientists to artists, to engage in a complex and wide-reaching debate about our impact on the planet in a time of rapid environmental change. Plants are important to consider, among many other environmental issues. Not only are plants a vital part to regulating our "emerald planet," but the human relationship with plants is also at a crossroads. Today, humans, more than natural forces, are the largest disperser of vascular plants on the planet.[15] The Wardian case played an important role in these dispersals.

Schutzglas—protective glass—was a word I remember from those days in Munich that continuously came up in our discussions with the exhibition designers and builders. The Wardian case is a good-sized object. The Berlin case is nearly four feet long and more than three feet high and wide. The exhibition builders wanted to know why they needed to put expensive protective glass around what appeared to be just a box. My fellow curators and I insisted that this was a rare object and in need of protection.

I have tried to track down other Wardian cases around the globe. To my surprise, only fifteen traveling Wardian cases exist. In Berlin there are two (a third is falling apart). At Kew Gardens there are eight—one from the nineteenth century and another seven from much later in the twentieth century (two of them are incomplete). There are two at the Hortus Botanicus of Leiden, both of which are still used in displays for visitors. There is one on a private estate in Tregothnan, in the south of England. There is one, which was only recently discovered, in Western Australia in the collection of the Waroona Historical Society; for much of its life it was used as a dog kennel. And there is one in the Smithsonian collection. Also worth noting is a beautiful eighteenth-century plant box, at the Jardin des Plantes in Paris, that predates the Wardian case. Hopefully more will be found in attics and garden sheds in the coming years. From a museum and historical perspective, the global rarity of such a significant object is extraordinary.[16] Why are there so few Wardian cases left in collections? The answer to this question reveals itself at the end of the Wardian cases' long and useful history.

I approach the Wardian case as both a historian and a curator. From a curator's point of view, this is a salvage study that brings together fragments of an object that has been all but lost. Curating, as a method, is about more than just objects; it is about the conversation between people, objects, and stories.[17] Curating also informs my practice as a historian. At most turns I follow specific Wardian cases, the motivations of the people using it, the plants on the move, and the outcomes of these transplants. It is to the beginnings of that story that I now turn.

PART 1

Possibilities

1

Experiments with Plants

In the summer of 1829 Nathaniel Bagshaw Ward prepared a wide-mouthed bottle for an experiment. Inside he placed soil, the pupa of a sphinx moth, and dry leaves, and then he fastened the lid of the bottle. It was one of his many curious experiments with the natural world. Inside the bottle, condensation formed on the glass and slid down the edges to the soil. He waited, watching every day, and finally the moth hatched. But upon observing the insect he saw two specks of green on the surface of the soil. After removing the moth, he resealed the bottle and placed it on the north facing window outside his study. The plants were a fern (*Dryopteris filix-mas*) and meadow grass (*Poa annua*). They lived untouched in the bottle for three years.

A medical doctor by training, Nathaniel Bagshaw Ward (fig. 1.1) inherited his father's practice at Whitechapel. From the surgery, it was a quarter mile's walk to his house at 7 Wellclose Square. The square was located up the road from the London docks, which in 1829 were the heart of British manufacturing, throbbing with life. Factories packed the vicinity around the docks; they operated constantly and belched their smoke into the air. Ward's house, like many others in the city reeling in the wake of the effects of the Industrial Revolution, was surrounded by smoke.

FIGURE 1.1 Portrait of Nathaniel Bagshaw Ward, 1859. Lithograph by R. J. Lane after the portrait by J. P. Knight. Courtesy Wellcome Collection, CC BY.

As a doctor and scientist he could see the impact of the pollution on his patients.

Ward's greatest passion was natural history. He spent his time away from the surgery collecting specimens and working his small garden. Before the bottle experiment, he built a rock wall at the back of the garden. In the woods surrounding London he collected ferns, mosses, primroses, and wood sorrels and then transplanted them onto the rock wall in the garden. At the top he placed a perforated pipe that let water trickle down the wall. Despite many attempts to keep his plants alive, they perished. And then he remembered his bottle experiment. One of the very ferns he had failed to grow on the wall had sprung to life inside the enclosed bottle. He thought about this fern and asked himself, "What were the conditions necessary for its growth?"[1]

The bottle created a micro-environment that Ward thought was free of soot, allowed light to enter, maintained a warm temperature, and was kept moist by means of the consistent cycling of condensation on the glass. Inside the bottle, the hardy fern produced four fronds annually, and the meadow grass, although seeds failed to form on it, flowered in the second year. The experiment only finished after the lid rusted, and rain-water entered, and both plants rotted. Undaunted, Ward replanted the bottle with a species of filmy fern (*Hymenophyllum*), which survived in the bottle for nine years. Following the success of the first experiment, Ward grew more than sixty different varieties of plants under glass in various bottles and cases.[2]

Ward's Wellclose Square home became a botanical laboratory in which experiments under glass were conducted. Outside the staircase window facing the north he built an eight-foot-square greenhouse he called the Tintern Abbey House. Inside the Tintern Abbey were at least fifty species of British and North American hardy ferns, as well as other plants such as camellias. He set out an "Alpine Case" with various high-altitude and mountain plants, including alpine azaleas, fairy primrose, alpine snowbell, and heath. Knowing there was not enough light for these plants, he placed the whole case on the roof of his house throughout the summer; all the plants flowered except an arctic bell heather. Then there was the drawing-room case, whose base held palms and ferns while suspended from the roof were small hanging pots planted with various cacti and aloes. He also experimented in these cases with bog plants in the base and succulents at the top. And in another covered glass case he experimented with an artificial gaslight during the night hours and dark cloth during the day to see if the plants survived; they did. To add color to his house he filled a case with spring flowers, such as Chinese primrose and Siberian squill, and put them on an outside windowsill in February; he had flowers for two months. He even built a greenhouse in his backyard, twenty-four feet by twelve feet and about eleven feet high, and manipulated the environment to accommodate plants from India and the Himalayas; he even introduced insects into it.

The house at Wellclose Square was a botanical wonderland. It was a product of his success with glass cases—plants abounded inside and outside the house. After visiting Wellclose Square, John Claudius Loudon, the leading botanical journalist of the day, described it as "the most extraordinary city garden we have ever beheld." Surrounding the house were thousands of plants in any space where they could fit. There were plants on top of all walls around the house, on top of the walls on the offices behind the house, covering the wall in the yard, along the cable ends, and on top of the lean-tos. Some plants were suited to the London environment, like the two figs that occupied the yard, or the sycamore that overshadowed them; but many were not, and these were placed under glass. Glass cases made to fit the house's many windowsills were installed next to all the windows on both the inside and outside. When Loudon visited Ward, he saw a glass case in the drawing room containing a magnificent blooming specimen of the South African giant honey flower (*Melianthus major*).[3]

The importance of what Ward achieved in his garden in polluted Wellclose Square cannot be overstated. Although the Wardian case was based on a simple idea, its effect was revolutionary. There were certainly some before him who had used glass cases effectively. But it was Ward's determination as an experimenter that compelled him to pursue and promote his ideas in a scientific and gardening society that was ready to listen. As Ward himself later reflected, "The simple circumstances which set me to work must have been presented to the eyes of horticulturalists thousands of times, but had passed unheeded in consequence of their disused closed frames being filled with weeds, instead of cucumbers and melons; and I am quite ready to confess, that if some groundsel or chickweed had sprung up in my bottle instead of the fern, it would have made no impression upon me."[4] Fortunately a small fern sparked his curiosity, and four years of experimentation brought evidence that Ward could report to the public. In 1833 the experiments were certainly not over. In fact, the biggest one was yet to come.

Atmosphere of Plants

Ward's greatest scientific contribution was the enclosed system. He showed that a plant in moist soil could be closed up under glass and placed in a location that receives light, in which state the plants could grow for many months, even years, without too much attention. Unlike a greenhouse, these cases were small and movable and required very little maintenance. Following the bottle experiment, Ward preferred to use a case to enclose his plants. They were generally made with a base of timber or iron and covered with a glass frame at the top. But they took many forms. Many were made to fit the size of the windows at Wellclose Square. The cases gave botanical flavor to the rooms and were constructed in various shapes and sizes: there was a simple plate with a bell glass over the top for a dining table, a small, ornate iron base with a glass top for the study desk, and a timber base with a hexagonal glass top for the sitting-room table.

Over the years the drawing-room case spread throughout Britain, Europe, and the United States as a necessary adornment for homes. The box that held the soil was usually made of well-seasoned mahogany one inch thick with a number of supports to hold the earth. At the top of the box were grooves to receive the glass cover, which was made of flattened crown glass (a type of optical glass used for lenses), except for a small door made of plate glass. The panes of glass were fitted together with putty and once assembled were painted to seal the glass structure. The case was usually four feet high (with legs), three feet long, and one and a half feet in breadth. These parlor cases were the precursors to modern terrariums.[5]

There were important scientific principles at work inside the cases. In Ward's house were living examples that many plants thrived in closed conditions, away from to the polluted London air. A closed case that could still admit light through glass gave the plants inside a different atmosphere from that outside. They themselves helped to create the environment they lived in: their respiration and photosynthesis (the process whereby plants convert light into energy) controlled the closed environment.

Many developments in understanding both plants and air appeared before 1829 that proved the logic of Ward's principle of gardening under glass. Developments in the field of plant physiology proved that plants could survive under glass in a closed system. In 1727 the British clergyman Stephen Hales found, using a "closed vessel," that not only do plants lose water through their leaves (transpiration), but they also use their leaves to draw "nourishment from the air."[6] In the 1770s the pioneering chemist Joseph Priestley famously noted "that plants are capable of perfectly restoring air." Later, in 1804, the Swiss chemist Nicolas-Theodore de Saussure laid the foundations for understanding photosynthesis by experimenting with plants in glass containers. He proved that for plants to grow they not only need water but carbon from the atmosphere so that they can use the energy from sunlight. By 1827 it was also proven that the pollution from factories in cities severely affected plants. In some places in Britain vegetation in the vicinity of a new factory died within two days of its opening. The atmosphere, as people were beginning to realize, is vital to plants, and plants are, in turn, vital to regulating the atmosphere.

When the world first heard about Ward's system of glazed cases in his letters to learned societies and journals, the simplicity of the idea captivated many leading scientists of the day, among them one of the most famous experimenters of all—Michael Faraday. In the 1830s Faraday was at the height of his powers and influence: he was in the process of publishing his research on electricity, which laid the foundations for it to become one of the most important technologies of the nineteenth century. Surprisingly, during this period Faraday found the time to deliver a lecture, on 6 April 1838, to the Royal Institution of Great Britain on the altered atmosphere created by Ward's glazed cases.[7]

In Edinburgh a group of scientists commenced a yearlong experiment to compare the air inside glazed cases to the "free atmosphere."[8] The results of the research proved the success of Ward's system. The main findings showed that the atmosphere inside the cases could be regulated by both the plants and the soil. They also showed that air expanded in conditions of light and contracted in those of darkness through the porous parts of the case's construction, such as the plug in the bottom of the case, the wood that formed the base of the case, and the putty between the glass plates. One of the authors concluded before the Botanical Society of Edinburgh: "Under the modified conditions with regard to climate, and the renovating processes in relation to water and air . . . the botanist and horticulturalist may be said to have entered on new and unexplored fields of vegetable research, and to have acquired the means of transporting to their own soil the varied and most delicate plants of every region on the earth." It was on this challenge of plant transport that Ward's next experiment was focused.

World's Greatest Garden

To transplant plants around the globe you needed to have connections. Nathaniel Ward had a pleasant and caring disposition, traits that served him well as a doctor and also allowed him to be surrounded by many friends as passionate about natural history as he was. He liked the company of others and enjoyed being part of a community of scientists, such as those at the Linnean Society and the Society of Apothecaries. It was these connections to many of the leading scientists and horticulturalists of the day that enabled his system of glazed cases to be widely recognized and quick to be adopted.

His most important friend in this early period was the nurseryman and scientist George Loddiges, of the famous Hackney nursery Loddiges & Sons. Loddiges worked tirelessly to continue the work of his father, Conrad Loddiges, who had set up the nursery in the late eighteenth century. By the time George took over the business, it had already been responsible for introducing many exotic plants to Britain. It was George's vision and passion, however, that made his family business at Hackney world-renowned. By 1836 Loddiges's nursery catalog listed a wide variety of plants in cultivation at their fifteen-acre nursery, including 67 varieties of oak, 29 of birch, 91 of crataegus, 180 of willows, 1,549 of roses, and 1,916 of orchids. They also had an impressive collection of palms and orchids. Loddiges & Sons were also the first to introduce bamboo to Europe. Loddiges's variety impressed visitors. As one garden journalist wrote after visiting the nursery, "In this department, Messrs. Loddiges have done more than all the royal and botanic gardens put together."[9] In the early nineteenth century Loddiges's nursery eclipsed all other royal or state-funded gardens; it was the greatest garden in the world.

Loddiges was also a visionary designer of hothouses, many of which were needed to keep exotic plants alive in London. By the beginning of the nineteenth century there was wide usage of both stove houses, which required continuous heating, and greenhouses which only required heating to avoid frost and extreme cold weather. By adopting new developments, such as better manufacture of cast-iron pipes, Loddiges was able to provide moist heat from a single furnace that was free of smoky and smelly fumes. He also developed a rain sprinkler to imitate tropical conditions and was able to keep alive the Mauritius fan palm, *Latania borbonica* (now *Latania lontaroides*), in one of his early hothouses.

On his property Loddiges went a step further and built one of the largest palm houses in the world. With a glass arch and held together by a frame of iron, it has been estimated to have been eighty feet long, sixty feet wide, and forty feet high. The palms inside came from Brazil, Ceylon, Egypt, Gambia, Jamaica, South Africa, St. Vincent, and Trinidad. In 1833 one journalist described the scene: "In the palm-house everything is in its usual luxuriance; the ferns are in most vigorous growth and the epiphytes flowering beautifully. . . . There is a beautiful new lycopodium (*L. circinatum*), the thick-set branches of which we can only compare to fine chenille work in embroidery." As well as the palm house, Loddiges had a camellia house and dry hothouse. The hothouses at Loddiges were the forerunners to the major glass structures that were built in Europe in the 1840s and 1850s, such as the Palm House at Kew Gardens and the Crystal Palace at Hyde Park. At the Loddiges nursery the developments in glass construction and the advantages of using glass in keeping exotic plants alive were obvious to visitors and customers.[10]

For Ward, George Loddiges was a good friend to have. Hackney was not just a pleasure ground; it was a business that imported, exported, and distributed plants. As well as running the business, Loddiges possessed the curiosity of a scientist. He was the leading authority on humming-birds, having the best collection in the world, and like Ward he was well connected with many of the leading societies of the day, including the Horticultural Society, the Linnean Society, and the Zoological Society.

When Ward told Loddiges of his discovery with growing plants sealed under glass, he was offered "kind assistance" and was given access to many plants at Hackney.[11] To Loddiges, Ward's experiments seemed revolutionary. Inside the cases were conditions similar to those in the wonderful hothouses at Hackney, yet with little need for additional heat or water.

Over the previous two centuries, one of the greatest challenges facing nurseries and botanical gardens was the successful movement of live plants from one country to another. These challenges had long hampered Loddiges's import-and-export business. If plants could grow successfully under glass, with little attention or care needed, then couldn't they be packaged into these special plant boxes and sent over the oceans?

At this time a short letter was read to the 4 June 1833 meeting of the Linnean Society of London announcing to the world Ward's experiments with plants. He told them of finding the fern and grass in the bottle; he told them how he placed the bottle outside his window for three years; he told them how he had experimented with other plants in different glazed cases. And then the glazed cases were unveiled: "I have the pleasure of submitting two of my boxes to the inspection of the Linnean society."[12] In the center of London in the summer of 1833, the glazed boxes were available to view by members of the Linnean Society and interested scientists.

Experiments over the Oceans

There were great challenges in bringing plants from distant continents. In the early 1830s the longest known sea voyage possible from London was to Sydney, in the distant British colony of New South Wales—a journey that took about six months, crossed from the Northern to the Southern Hemisphere, and passed through many different climates. In June 1833, around the same time as the Linnean Society was learning about the glazed cases, Ward and Loddiges packed two cases with a collection of ferns, mosses, and grasses and delivered them to their friend the sea captain Charles Mallard, at the St. Katharine Docks at the Port of London, who was preparing to leave for Sydney.[13]

On 8 July 1833 the barque *Persian*, commanded by Mallard, set off down the Thames for the British colonies in Australia. The *Persian* was loaded with 394 tons of cargo and fifty-seven passengers, of whom forty-four were in steerage. Taking pride of place on the spacious poop with the helmsman were the two glazed cases. The poop deck was positioned high on the rear of the ship and afforded the helmsman and captain an elevated position from which to navigate and view the sails. For a glass case full of plants, it was also the location on the ship that received the most light and was most protected from salt spray. Mallard sailed south and round the Cape of Good Hope on his way to Hobart, Van Diemen's Land, the small island hanging off the southeast of the Australian continent.[14]

While awaiting news from Mallard, Ward prepared another case to show off his experiments. In December 1833, as London entered winter, he put one of his glazed cases on display at the meeting rooms of the Society for the Encouragement of Arts, Manufactures and Commerce (RSA), one of London's many societies that emerged at the end of the Enlightenment with an interest in moving plants—an interest that extended over at least a century. Planted in May, many months earlier, Ward's fern-filled case was available for the Society's many patrons to view. The box was laid with broken bricks, sand, and sphagnum moss for drainage, over which the ferns were planted in topsoil (what nineteenth-century gardeners called "vegetable mold"). They were well watered, and the water was allowed to drain out of the hole in the bottom of the box; the hole was then plugged and the glazed lid was placed on top (fig. 1.2). As Ward instructed the Society's members, "No further care is required than that of placing the box in the light." He went on enthusiastically to members of the Society: "Many other plants . . . I had previously attempted in vain to grow in town, succeed equally well under this plan of treatment." He concluded by lamenting the "deteriorating influence to town air" that was causing many plants to die; in his cases, however, many different varieties could be grown for Londoners' enjoyment.[15]

FIGURE 1.2 Traveling-style Wardian case, as described by Nathaniel Ward. From N. B. Ward, *On the Growth of Plants in Closely Glazed Cases* (London: John Van Voorst, 1852).

Meanwhile, at sea his ferns and grasses were sailing over the southern oceans. The *Persian* docked at Hobart five months after departing London. After the passengers disembarked and the cargo was unloaded, Mallard had the opportunity to assess the experiment. While he dashed off a letter to an anxious Ward, the two glazed cases remained on the poop deck where they had been throughout the entire voyage, still awaiting their final destination: Sydney. Mallard began his 23 November 1833 letter: "You will, I am sure, be much pleased to hear that your experiment for the preservation of plants alive, without the necessity of water or open exposure to the air, has fully succeeded." Only three ferns died on the voyage, Mallard reported; the rest were doing well, with the thriving grasses pushing against the top of the box. Mallard concluded the letter to Ward with "warm congratulations upon the success of this simple but beautiful discovery for the preservation of plants in the living state upon the longest voyages."[16]

The *Persian* arrived in Sydney on 2 January 1834. Fifty-two passengers disembarked, most of them having taken the cheapest fare possible in steerage, and most of them on the ship only for the short eleven-day voyage from Hobart. Then the ship was unloaded. Among the goods were 200 boxes of soap, 62 casks of salt, 48 cases of felt, 18 barrels of varnish, 7 cases of hats, 440 bars of iron, 5 bundles of steel, 218 tierces (equivalent to the size of an oil barrel) of beer, 5 casks of wine, 4 cases of screws, 2 pianos, 16 boxes of tea, and 365 pieces of machinery. And, of course, two glazed boxes of plants from London that had been opened only once in seven months.[17]

Ward was friends with both of the well-known plant explorers Allan Cunningham and his brother Richard. Following his explorations in Australia in 1817–31, Allan returned to London and told Ward of the arid southern continent that abounded in new plants, which he had witnessed on his explorations. Meanwhile, Richard became the superintendent of the Sydney Botanic Garden. It was this connection, as well as that of Captain Mallard, that enabled the experiment of sending plants in glazed boxes. Richard was due to receive Ward's two cases of plants. But by the time Mallard arrived in early 1834, Richard was in New Zealand searching the kauri pine forests for trees that would make suitable topmasts. His assistant, John McLean, received the cases. Inside the case nearly all of the plants were "alive and flourishing."[18]

From Sydney, Mallard wrote to Ward detailing the successful transfer of plants to the botanical gardens. He concluded his letter by assuring Ward that "the complete success of your interesting experiment has been decidedly proved." But there was still the return journey.

In February McLean filled the glazed cases with plants collected from the Sydney region, and delivered them back to the *Persian*. When the ship, laden with colonial produce, sailed in late May 1834, the temperature hovered between 90 and 100 degrees Fahrenheit. The two glazed cases were again placed on the poop. By the time the *Persian* rounded Cape Horn, sailing east, there was a foot of snow covering the deck, and the temperature had fallen to 20 degrees. When the ship called at Rio de Janeiro the temperature was on the rise, reaching nearly 100 degrees, and it continued to rise as the vessel crossed the equator. By the time the *Persian* arrived at the English Channel the temperature was 40 degrees—

barely above freezing. The fluctuation in temperature was a considerable test for the plants inside the cases.[19]

Six months after leaving Sydney, in November 1834, the *Persian* landed in London. Upon hearing that the ship had arrived, both Ward and Loddiges made their way to the docks. When they boarded the ship and looked inside, Loddiges was overwhelmed with delight. The plants "were in the most healthy and vigorous condition."[20] Among the ferns they found healthy fronds of the delicate coral fern (*Gleichenia microphylla*), an Australian plant never before seen in Britain. The coral fern would later be drawn by Franz Bauer and find its way into William Hooker's *Genera Filicum* (1842), one of the most important early books published on ferns. Along the journey Australian black wattle seedlings had accidentally sprung up from seed in the Sydney soil. The experiment was a success.

Modernizing Egypt

Buoyed by the shipment to Australia with Captain Mallard, Nathaniel Ward continued moving plants in glazed boxes. This time the boxes were used with greater confidence—more boxes with wider varieties of plants. In the summer of 1834, after he received word of Mallard's outgoing success but before it had arrived back in London, Ward was asked to assemble a selection of plants destined for Egypt and Syria. The request came from Ibrahim, pasha of Egypt, the successful general of the newly formed Egyptian army and son of the Egyptian ruler Muhammad Ali. Ibrahim had palaces and gardens on the island of Roda and on the eastern bank of the Nile, and also in newly acquired territories in Damascus and Beirut.

A few years earlier James Traill, a young undergardener from the Horticultural Society of London, had taken up an appointment as the pasha's gardener. On his outward journey Traill took a collection of plants from the Society, but they perished during the voyage. By the time Traill became head gardener in Egypt in the 1830s, many of the gardening elite in London, including the Horticultural Society, were beginning to hear of Ward's success. The pasha's London agents contacted Ward to ask him to supply new plants for his gardens. Ward accepted and used his new method to transport them.[21]

Six glazed boxes containing 173 species were sent. It was a diverse selection. There were ornamentals, such as zebra plants and orange trumpet creepers; trees, such as rubber trees and Florida royal palms; and vines, such as moonseed. There were medicinal plants, including betel, dwarf cardamom, manaca, and spices, such as allspice, black pepper, Chinese cinnamon, ginger, turmeric, and vanilla. There were useful trees such as bombax cotton, candlenut, Spanish cedar, custard apple, old fustic and sago palm. Most surprising in this collection was the diversity of species that Ward had compiled. Unlike the ferns and grasses used in the Australia experiment, the plants destined for Egypt were challenging to move. This was another big step in his experimentation. Considering that he was sending them to such a high-ranking member of Egypt, he must have had confidence that such a diverse array of plants could survive the journey.[22]

The diverse plants were destined for the upper classes in the emerging modern Egypt. When the paddle steamer *Nile*, a ship specially prepared at Fletcher & Fearnell's Blackwall shipyard for the Egyptian navy, sailed in early August 1834, it carried not only Ward's glazed cases but also men to serve in Egypt. Commanded by William Light, who was friends with the Egyptian ruler and his son and was in England recruiting British officers for the Egyptians, the *Nile* was due to join the Egyptian fleet following the invasion of the Ottoman lands in Syria. The journey from England took two months. Of the six boxes, two were destined for Traill, the pasha's gardener, outside Cairo, and four for the gardens in Syria, one at Damascus and the others at Beirut (then still part of Syria). The pasha was governor-general of these areas and in Beirut had assumed command at the elevated position on the Grand Serial, "the barracks," and wanted plants transported to the gardens there. Following the Second Egyptian-Ottoman War (1839–41), which forced the pasha's retreat from Beirut, the British Colonel Gordon Higgins saw a number of Ward's glazed cases still in the gardens.[23]

Ward anxiously awaited word of the plants' conditions. Traill wrote from Cairo after receiving the plants in a special consignment from Alexandria, "The collection was received here in the very best condition: the plants, when removed from the cases, did not appear to have suffered in the slightest degree; they were in a perfectly fresh and vigorous state, and, in fact, hardly a leaf had been lost during their passage."[24] Traill concluded his letter to Ward: "Your plan, I think decidedly a good one, and ought to be made generally known." Following this success, a case made up solely of coffee plants was dispatched with the same success. Many of the plants that Ward sent to Traill acclimatized to the Egyptian environs very well, particularly the custard apple, turmeric, arrowroot, and ginger. Some of the trees had grown dramatically. In four years the Spanish cedar grew to over fourteen feet.

Within a few short years of its invention, Ward's glazed plant case had made two significant journeys and reached through a circuitry that was impressive. Starting with Ward and then his friend Loddiges at his nursery in Hackney and spreading through a small circle of botanists and gardeners in London circles, the glazed case quickly spread beyond Britain to Australia and Egypt. With the spread of European empires, the early success of Ward's invention made it a valuable tool for anyone with a keen interest in plants. The possibilities of moving both horticultural novelties and agricultural necessities across the oceans from one country to another now appeared almost limitless.

2

....

A Brief History of the Plant Box

A plant with its flowers, fades and dies immediately, if exposed to the air without having its roots immersed in a humid soil, from which it may draw a sufficient quantity of moisture to supply that which exhales from its substance, and is carried off continually by the air. Perhaps, however, if it were buried in quicksilver [mercury], it might preserve for a considerable space of time its vegetable life, its smell and colour. If this be the case, it might prove a commodious method of transporting from distant countries those delicate plants which are unable to sustain the inclemency of the weather at sea, and which require particular care and attention. —BENJAMIN FRANKLIN, letter to Barbeu Duborg, 1773*

Nathaniel Bagshaw Ward was not the first to move plants across the oceans. The impulse is as old as the human passion for travel. "Canoe plants" is the wonderful name botanists have given to plants that were moved by the great Polynesian navigators who traversed the Pacific; as they followed the stars to distant locations, they took with them plants from different islands. As early as 1450 BCE the Egyptians moved live incense trees from Somalia, as shown in reliefs on the Deir el-Bahari mortuary temple. When cultures set out to explore new lands, they loaded their boats with their own plants and animals. Often the plants were useful ones, but sometimes they were simply things of beauty that reminded them of home. It was not as simple as carrying a few seeds along for the journey. Some seeds are very difficult to find and move. Seeds might not be available at the time of travel, or they might be oily and quickly spoil, or they might have a short life span. For these reasons, it was often live plants that had to make the journey.

For Europeans, plants were on the move as early as the Enlightenment, when people's curiosity swelled beyond the borders of their own nations. A plant was a delicate traveling companion. Fresh water was at a premium on a long ocean voyage, and thirsty plants were certainly not the first in line to get a drink. Moving live plants to distant locations was a great challenge, and scientists tried many methods. Benjamin Franklin even pondered burying them in mercury.[1]

For many years the box, simple and square, was the common method for moving plants. Centuries before Ward proposed the glazed case, many different varieties of boxes were used for this purpose. Following the exploration of new worlds in the fifteenth and sixteenth centuries, the seventeenth century saw the expansion of European centers of trade and imperial activity. During the Renaissance a new program of science, at the time called "natural philosophy," gradually emerged. One aspect of the larger program, originally laid out by Francis Bacon, was an endorsement for travelers and voyagers not only to witness exotic lands, but also to gather their rarities and curiosities in cabinets and display them as objects of interest.[2] For Europeans, moving plants across oceans and around the globe increased steadily throughout the seventeenth century.

New and curious plants came from the colonies in the Americas, the East Indies, the West Indies, and the South Seas. As the seventeenth century progressed, valuable plants were moved greater distances and with more regularity as trade and shipping increased. With science and gardening gathering momentum as serious pursuits in Europe, the thirst for more interesting plants, and knowledge about them, increased. But many different varieties of plants could not survive the journey from the New World or the East.

Working out the best method for transplanting useful plants became a question that occupied natural historians. Should a covered wooden box laid with soil be sent? Or was it best to follow the system used by the East India Company and use a cask with holes bored in the side? Or should a pot be used? All these methods were tried by naturalists. Lasting almost two centuries, from the middle of the seventeenth century to the early nineteenth century, their efforts paved the way for the Wardian case.

One of the earliest reports describing the study and collection of natural specimens was Robert Boyle's advice in 1666, *General Heads for the Natural History of a Country, Great or Small: Drawn Out for the Uses of Travellers and Navigators*. In it he outlined a simple approach for the study of nature whereby collection and observation could be undertaken by those untrained in natural history, such as ship captains, merchants, and colonial travelers. Boyle described many natural objects that were of interest to European collectors and could be found while traveling in Turkey, Egypt, Guainía, Poland, Hungary, Brazil, Virginia, and the Caribbean Islands.[3]

Following Boyle, Hans Sloane was one of the most well-known naturalists of the seventeenth century and a keen mover of plants. Among Sloane's many well-known exploits, he became president of the London Royal Society (1727–41) after Isaac Newton and was also instrumental in setting aside the Chelsea Apothecary Garden in London (later known as the Chelsea Physic Garden). Both of these institutions were critical in establishing the value of foreign plants and the need to move them. Sloane traveled to Jamaica in 1687 and collected widely, arriving back in London with some of the first herbarium specimens to arrive in Britain. When Sloane died, his large personal collection formed the basis of the collection at the British Museum of Natural History.

Even before Sloane left for Jamaica, he was familiar with the cacao plant, one correspondent to the Royal Society wrote from Jamaica in 1673: "I Send you on this Ship a box, that hath in it a Cacao tree painted to the life. 'Tis certain, nothing was ever more like." Already in the late seventeenth century boxes for plants, even plant replicas, were being sent across oceans to keen collectors. While herbarium specimens and painted impressions were important, getting live specimens across oceans was a great challenge. As Sloane's extensive correspondence shows, in the late seventeenth century plant collectors from many countries were in communication with one another about the best method of finding and sending plants. Even with this interest, by the late seventeenth century only about five hundred new plants had been introduced into Britain, many of these hardy trees from America.[4]

The voyage across the Atlantic from America to England was relatively short, making for greater success in sending live plants. In the 1690s John Evelyn used his friendship with Samuel Pepys, the secretary of the Admiralty Board, to arrange for plants to be brought back on His Majesty's ships returning from New England. Evelyn told senders that for journeys longer than eight days, the roots of plants should be coated with honey to help preserve them.[5] In 1696 another naturalist, John Woodward, presented to the Royal Society his *Brief Instructions for Making Observations in All Parts of the World; as also for Collecting, Preserving and Sending Over Natural Things.* His instructions, as the title suggested, were intended for all parts of the world, particularly Africa and the West Indies. Woodward set out specific instructions for fern transport: the roots should be "wrapt up in a lump of *Clay or Loame* and then put in a Box with Moss, and so sent over."[6] Boyle's and Woodward's were some of the first such instructions for plant collectors.

James Petiver was an apothecaries' tradesman who dispensed medicines and offered descriptions of natural history curiosities from all over the globe. By the early years of the eighteenth century he had built up a correspondence and supply network that included many in North America and the West Indies. With growing concern about the condition of specimens that he received from his suppliers, he printed a broadsheet titled *Brief Directions for the Easie Making and Preserving Collections of all Natural Curiosities for James Petiver Fellow of the Royal Society London.* In it he noted that the best way for plants to travel was as either seeds or flowers. But knowing that this was not always possible, he suggested simply, "Then gather it as it is."[7] Many of the species that Petiver compiled in his books were provided by commercial travelers.

Bartram's Boxes

Fascination with foreign specimens of natural history increased in the eighteenth century, with many of the emerging merchant class using their global commercial connections to import plants from elsewhere. One very successful network, between London and North America, was led by the London cloth merchant Peter Collinson and the Philadelphia farmer and collector John Bartram. Bartram sent boxes of seeds and hardy plants from Philadelphia to London. Very early in their relationship, Collinson described to Bartram the best box to send: "A great many may be putt in a Box 20 Inches or Two feet square and 15 or 16 inches High & a foot in Earth is Enough. . . . Nail a few small Narrow Laths across it to keep the Catts from Scratching It."[8] Collinson clearly shows the problems that plants faced on the ship: they carried a range of travelers, among them cats, rats, and livestock, that yearned for fresh soil and foliage.

A box similar to that prescribed by Collinson contained some of the earliest successful arrivals in London. Collinson wrote Bartram in January 1735: "Thee canst not think how well the little case of plants came, being put under the captain's bed, and saw not the light till I went for it; but then, Captain Wright had a very quick passage."[9] Other boxes arrived in a terrible state. On one occasion a box of plants arrived in London in which the only things alive were a family of pink rats.[10] Later John Fothergill, another collector, certainly aware of Collinson's experience, recommended breaking glass and putting it in the soil to deter rodents.

The eighteenth-century plant trade between North America and Britain was an important step in working out how to move plants successfully. As the demand for new plants grew, they devised an inventive scheme of subscribers to "Bartram's boxes." Bartram would send a wooden crate to each subscriber. Each cost five guineas and usually contained around a hundred varieties of seeds, dried plant specimens, and other natural curiosities. Bartram's subscribers included lords, bishops, natural historians, and nurseries. Some of Bartram's introductions to England included magnolias, azaleas, rhododendrons, mountain laurels, sumacs, and sugar maples.[11]

Although gardening took off during the eighteenth century, there were still many challenges in sending live plants across the ocean. Not only was it difficult to keep them alive during the Atlantic crossing; other factors also played a part. During the Seven Years' War (1756–63) the capture of British ships transporting Bartram's boxes was one cause for loss. As the tensions of war escalated, Bartram decided to split his packages among various ships. He also addressed his packages not only to his English subscribers but also to noted botanical men in France and the Netherlands, such as Thomas-François Dalibard; Georges-Louis Leclerc, comte de Buffon; Bernard de Jussieu; and Jan Frederik Gronovius. If the French captured the ships and saw them addressed to these noted botanical men of science, then possibly Bartram's collections and notes would make it through. As Bartram said, he preferred that his boxes fall into "learned, [rather] than ignorant hands."[12]

In 1752 Henri-Louis Duhamel du Monceau, a French naturalist and member of the Académie Royale des Sciences, published an important set of instructions for travelers who intended to collect plants: "Those who, for their utility, or to satisfy their tastes or those of others, want to transport plants or seeds, or other natural curiosities a great distance must know that these transports are almost always a pure loss, for lack of necessary precautions."[13] The challenges of moving plants were widespread. As early as 1723 another Frenchman, Gabriel de Clieu, on a trip from Paris to Martinique, took with him coffee plants cut from specimens in Paris's Jardin du Roi. He carried them in a small box and on the journey had to share his very limited water ration with the thirsty coffee plants. After the journey de Clieu remembered the "infinite cares that I had to give to this delicate plant during the long crossing."[14] The coffee survived and would be the start of coffee plantations in the Caribbean islands of the New World.

The delicate plants required dedicated care to survive. In the early eighteenth century Kew Gardens had special hothouses for nursing sickly plants back to health after long sea voyages. Indeed, the island of St. Helena became a halfway point for not only sick sailors, but also for ailing plants on the long voyage back to England.[15]

Around 1770 another well-known naturalist, John Fothergill, wrote instructions for his collectors. *Directions for Taking Up Plants and Shrubs, and Conveying Them by Sea* was a single page that accompanied his correspondence with collectors and ship captains going to North America. Fothergill was one of London's leading physicians as well as a keen horticulturalist. He recommended that young plants about a foot high were the best to collect. Fothergill suggested boxes that were four feet long, two feet deep, and two feet wide for long journeys. When half filled these could be handled by two men carrying them on board a ship. Over each of the boxes hoops with small ropes twisted between them to form a net were used to cover the tallest plants. The net prevented dogs and cats from disrupting the plants. Each box, hooped and netted, also had a canvas cover that could be used in bad weather. The captain who took charge of the plants had to be "particularly informed that the chief Danger Plants are liable to in Sea Voyages is occasioned by the minute Particles of Salt Water with which the Air is charged, whenever the Waves have the white frothy Curls upon them." Fothergill also warned against shutting them up all the time owing to the "Stagnation of the Air under the Covers" (fig. 2.1).[16]

Ellis, Thouin, and the Plant Box

In the eighteenth century, as the centers of Europe enjoyed greater access to new and exotic objects, it was one of the most widely remembered naturalists of all, the Swede Carl Linnaeus, that outlined a system for classifying all these objects of nature. First with the short work *Systema Natura* (1735) and later in *Species Plantarum* (1753), Linnaeus described his taxonomic arrangement of nature, whereby plants and animals were classified using their means of sexual reproduction. At its core was the binomial naming: each plant was labeled according to a species and genus. This replaced the earlier polynomial practice, in which plants received long descriptive names. Linnaeus's system allowed for all living things to be labeled using two Latin names, enabling those names to be both memorable and universal.[17] Linnaeus required an enormous library of specimens to develop his system and was therefore a prolific collector and correspondent.

In England, John Ellis was one of the key champions of Linnaeus's new classification system and a great supplier of specimens to the Swede. Ellis maintained a close correspondence with Linnaeus over many years. Ellis is widely remembered as an expert on zoophytes—animals

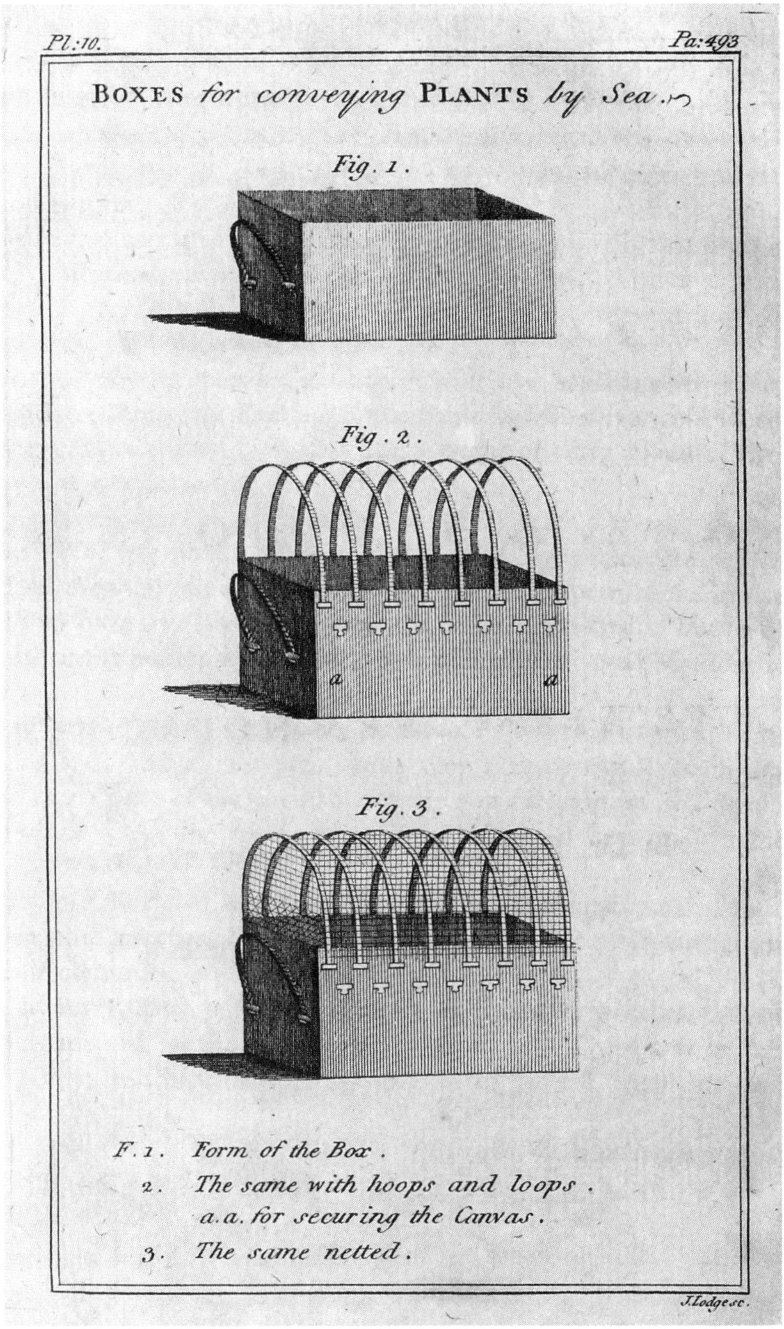

FIGURE 2.1 John Fothergill's boxes for transporting plants by sea, ca. 1770. From *The Works of John Fothergill* (London: Charles Dilly, 1784).

that resemble a plant, such as sea anemones, sponges, and corals. His pioneering work brought him into close contact with many leading botanists of the day, including not only Linnaeus but also Fothergill and Collinson. Late in his life, noticing the unsuccessful attempts of many of his correspondents, Ellis committed much time to solving the challenges of moving plants, focusing not only on importing exotics to England but also on moving them between the colonies. In addition to his other work, Ellis was the British agent for Florida and Dominica, a position that brought him many contacts in the colonial office and in distant locations.[18]

Ellis's numerous publications in the *Philosophical Transactions of the Royal Society of London* show that he was publishing on plant transport as early as the 1750s.[19] In 1770 he published a pamphlet titled *Directions for Bringing Over Seeds and Plants from the East-Indies and Other Distant Countries in a State of Vegetation*. Over the previous decade Ellis had made many short reports on his various experiments with moving plants and seeds, but the pamphlet was intended to reach a wider circulation among his correspondents in Europe, India, China, and the Americas. The survival rate of foreign plants and seeds was low, Ellis wrote: "It might be reasonably supposed, from the great quantity and variety of seeds which we yearly receive from China, that we should soon be in possession of the most valuable plants of that vast empire; yet it is certain, that scarce one in fifty ever comes to any thing, except a few varieties of annual plants, which have been common in our gardens for many years."[20]

Showcased in Ellis's *Directions* were the best boxes for transporting seedlings and live plants. The boxes were three feet long, fifteen inches wide, and eighteen to twenty inches deep, with a "cover of wire" to deter vermin and a hinged lid. The sides needed numerous small holes near the top to allow air circulation—"to let the crude vapours pass off that arise while the cover is obliged to be let down."[21] For Ellis, "air is absolutely necessary," and one must always "let out the foul air."[22] If a box could not be made, then a cask could also be effective. At the bottom of either box or cask there should be a layer of wet moss, or, if that was not available, decaying wood, to provide moisture and drainage. On top of this was soil and then another layer of moss. He also noted that it was important for the traveler to save as much rainwater as possible to supply the plants. Ellis concluded that by successfully moving plants curious gentlemen, intelligent seedsmen, and gardeners would "both do honour to themselves, and a real service to their country."[23]

Over two decades John Ellis devoted a large amount of time to experimenting with different types of cases for carrying plants. The box was one of the main methods he promoted. Ellis's were some of the first boxes designed specifically for moving plants and were certainly some of the first to appear fully sketched in print for the purpose of encouraging voyagers to use similar boxes specific to the task of moving plants (fig. 2.2). It was Ellis's designs and recommendations that were copied and followed well into the nineteenth century.

Shortly before his death Ellis published a pamphlet titled *A Description of the Mangostan and the Bread-Fruit* (1775). In addition to proclaiming the importance of each of these plants, the former "the most delicious" and the latter "the most useful," he included directions to voyagers for bringing over the breadfruit tree. Leafing to the middle of Ellis's book, one can find his designs for another type of box: a wired case. Presciently, the wired case that Ellis designed could also be fitted with glass, "by which means the plants may receive the benefit of the sun, and at the same time be preserved from the severity of the cold."[24] The wired cases were made, following Ellis's designs, by the carpenter John Burnham operating out of Brooks-market, Holborn. In the 1770s Ellis showed these wired cases at John Gordon & Co.'s Seed Shop on Fenchurch Street, not far from the Port of London. Well-known as one of the most important shops for London plantsmen, it was an important venue for displaying the value of using boxes to transport plants. As well as the wire cases for the breadfruit, two other designs of Ellis's could be seen at Gordon's shop.[25] In his books he provided drawings of all cases so that "such gentlemen who go abroad with a resolution to promote the object of this address may be furnished with a convenient apparatus, for transmitting either these or any other useful or rare plant, to Great Britain."[26]

Throughout the eighteenth century, the simple box or case for mov-

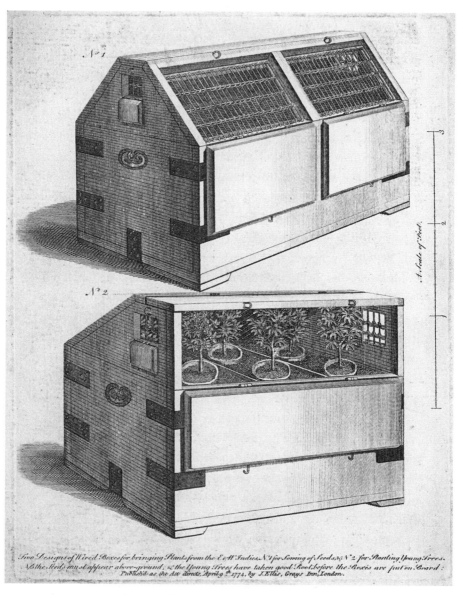

FIGURE 2.2 Two plant boxes designed by John Ellis, 1774. From John Ellis, *A Description of the Mangostan and the Bread-Fruit* (London: Edward and Charles Dilly, 1775).

ing plants and seeds was well suited to the context of British collecting. Naturalists in England relied mostly on commercial networks and personal relationships to obtain plants from distant locations. Across the channel, however, French scientists enjoyed a much greater level of state support. While the British relied on routes tracked by transatlantic commerce, the French used extensive connections within the French crown to move plants into Paris.[27] Their collections (mostly from across the Atlantic) were transported mainly on government vessels.

In the mid-eighteenth century France's André Thouin was a leading figure in the movement of plants around the globe, and he saw the value of the use of boxes for the purpose.[28] The head gardener at France's premier garden, the Jardin des Plantes, he was widely connected on the Continent. His close collaborators included Guillaume-Chrétien de Lamoignon de Malesherbes, Jean-Jacques Rousseau, and Augustin Pyrame de Candolle. He was a member of the French Academy of Sciences and regularly exchanged plants with foreign botanists. Thouin's plant box, still preserved at the French National Museum of Natural History, is the oldest known plant box surviving today. Thouin was also influential in recommending Jean-François de Galaup, comte de Lapérouse, to take plant boxes on his expedition. In 1824 Thouin fell gravely ill, and his major work, *Cours de culture et de naturalisation des végétaux* (1827), was completed by his nephew Oscar Leclerc; it showcased many of the plant boxes that French botanists used, with Thouin's promotion, in the previous century. However, in the late eighteenth century things had started to change.

Expanding Empires

The late eighteenth century marked a turning point in the transport of plants. Europe, with many expanding colonies, and a widely developing interest in exotics was welcoming many new species of plants. Of all the plants growing in 1789 at Kew Gardens, nearly one-third were exotics.[29] Describing nature with herbarium specimens was no longer enough: being able to see the plant growing in its natural environment became a serious pursuit for many gardeners and horticulturalists, and provided the upper class with objects of aesthetic appreciation. Joseph Banks's vision was much broader than the boxes that John Ellis promoted. For Banks, the entire ship became a means of moving botanical specimens.[30]

As a young man Banks was on Captain James Cook's *Endeavour* as it sailed around the world. Because he was sailing as a private person paying his own way on the voyage, he was able to have the great cabin remodeled in order better to store natural specimens. Interestingly, independent of Banks, Cook had consulted the seedsman John Gordon from Fenchurch Street and had taken with him seeds of useful plants hermetically sealed in glass bottles.[31] Cook subsequently planted the seeds in a garden in Tahiti. Upon Banks's return from the *Endeavour* voyage, his influence and connections grew, and these connections helped him to carry out many of his plans for new ways of moving plants. Banks used his prodigious botanical knowledge and familiarity with the challenges attendant on sending live plants in two very important government expeditions in the late eighteenth century: the transplantation of the breadfruit tree to the West Indies and the transport of useful plants to the colony of New South Wales.

To move the breadfruit tree, Banks repurposed the *Bounty*. It was like no other ship commissioned in His Majesty's Navy. Below deck, when it left London in December 1787 on its thirty-thousand-mile voyage, it was a purpose-built plant transporter. The captain's cabin was taken over by pots for trees, and an elaborate watering system was used to sustain the trees.[32] The story of transplanting the breadfruit tree is one of the most infamous but also one of the most instructive on the complexities of moving plants. In particular, the desire to move plants always had built into it elements of imperial global reach and economic interests. Banks was in collaboration with William Pitt, the British prime minister; Henry Dundas, head of the India Board; and Lord Sydney, the Secretary of State, in trying to enable the West Indies to supply a higher grade of cotton. Under instructions from Banks, a young botanist was sent to India to obtain seeds of a better variety of cotton, which were to be taken to the West Indies. But more cotton in the West Indies to supply the mills in Britain required more slaves, which required greater food production. Therefore, the high-yield breadfruit was to be transplanted to the West Indies. The idea for the transport was not new; many had hinted at it in the 1770s. But it was not until the *Bounty* sailed in 1787 that the grand plan was realized.

No transplant had ever had so much state investment as the bread-fruit tree. "The ship is to be fitted at Deptford," wrote Banks, "for the greater security of the Trees and Plants which she is to take on board."[33] The great cabin was completely refitted to hold breadfruit trees. But after leaving Tahiti with a boatload of trees, the crew mutinied, and Captain Bligh, along with his gardener and other loyalists, were put off in a life-boat. The loss of the ship was initially a huge failure. Less well-known is that a few years later, in 1792, with Bligh again at the helm, the *Providence* was fitted for transporting trees in the same way as the *Bounty*. About two-thirds of the plants died in transit, but approximately 690 plants reached the West Indies. For the long history of the plant box, the story of the breadfruit shows that the challenge of moving plants was so great that whole ships were appropriated for the purpose.

After the breadfruit transplant, Banks continued to use grand ways to move plants, importantly to supply useful plants to the newly established colony in Australia.[34] At around the same time as the *Bounty*, the First Fleet of eleven British ships carrying fourteen hundred convicts sailed for Botany Bay in 1787. From the outset Banks was involved in equipping these ships with plants and seeds that could help establish the colony. The first few years did not play out well, and a well-stocked supply ship, the *Guardian*, was quickly fitted out in England for the purpose of re-plenishing supplies. Among the things that were in high demand were plants. In the summer of 1789, while the *Guardian* was docked at Wool-wich being loaded with supplies, Banks made a special trip to see the vessel for himself. He wanted to refit the great cabin, as he had done with the *Bounty*, with enough space for ninety-three pots, but the ship was already reaching capacity. Banks devised another plan: a greenhouse.[35] Working on board with the master of the ship and the shipbuilder at Woolwich, Banks proceeded to chalk out an area fourteen and a half feet long and almost twelve feet wide for the plant cabin. He instructed that it should be five feet high, positioned on the quarterdeck, and built within eight days. Once built, the cabin was home both to the plants and to the gardener John Smith. Not only was the plant cabin to take plants to the new colony, but upon its return it would return with rare native Australian plants for Kew Gardens.

Drawing upon knowledge accumulated over the past century, Banks issued many handwritten notes to gardeners who were accompanying shipments across the oceans. Many of his recommendations are similar to those instructions provided by other gardeners and naturalists. He warned against salt spray and also against any pests that could get at the plants: cats, cockroaches, monkeys, and rats were all specified. While pests were to be excluded, fresh air was to be admitted whenever possible. Ultimately, what Banks required of the gardener was commitment: "Plants on board a Ship, like Cucumbers in February, require a constant attendance."[36]

The *Guardian* never made it to the new colony in Australia; after rounding the Cape of Good Hope, it struck an iceberg and was eventually wrecked off the coast. The first thing to go overboard when disaster struck was the heavy and cumbersome plant cabin.[37] Despite this failure, the lasting influence of Banks's idea can be seen in the many ships that followed. The replacement cargo, sailing aboard the *Porpoise* in 1800, also had a plant cabin built on the quarterdeck.[38] Between 1787 and 1806 Banks was instrumental in seven ships' having plant cabins placed on the quarterdeck.[39] The plant cabin was an expensive addition to many ships and was phased out after a decade of use. By the time of Banks's death in 1825, using plant boxes and cases for transport on other vessels was by far the most common means for transporting plants.

Back to the Box

As shipping increased and the world became increasingly connected through exploration and trade, even in the early nineteenth century transporting live plants still faced great challenges. In 1819 John Livingstone, the keen botanist and surgeon posted in Macao for the East India Company and a corresponding member of the London Horticultural Society, wrote on the challenge of sending live plants from China to London. Livingstone estimated that only one in a thousand plants survived the journey. If his calculations were correct, given that the average cost of purchasing plants in Canton (including their chests) was six shillings and eightpence, then every plant that had now made its way into England must have cost more than £300 per plant. He proposed a number of plans for the successful movement of plants. One was simply to send a gardener with every dispatch of live plants. But whatever the method, Livingston concluded, it "becomes a matter of importance to attempt some more certain method gratifying the English horticulturalist and botanist, with the plants of China."[40]

John Lindley was an important figure in nineteenth-century botany and the movement of plants. With his energy and productivity, he was a leading member of the London Horticultural Society and would go on to become the first professor of botany at the University College, London. Lindley had been with the Society for just two years as the assistant secretary when in 1824 he bemoaned the great challenge in sending live plants across oceans: "The idea which seems to exist, that to tear a plant from its native soil, to plant it in fresh earth, to fasten it in a wooden case, and to put it on board a vessel under the care of some officer, is sufficient, is of all others the most erroneous, and has led to the most ruinous consequences."[41]

In the early nineteenth century people were trying many methods for moving live plants, and Lindley received many of these boxes at the Horticultural Society's Chiswick gardens. In the 1820s one variety of case successfully came from China, sent from the East India Company employee John Reeves, who was based in Macao. Reeves's portable greenhouses predated Ward's designs. As the *Gardeners' Chronicle* described the boxes, "Not a company's ship at the time sailed for Europe without her decks being decorated with the little portable greenhouses." Other interesting boxes came from the islands of the Indian Ocean. Accompanying Lindley's paper to the Horticultural Society were designs for a glazed box that he particularly liked that had been sent by the governor of Mauritius, Robert Farquhar. This box, three feet long and four feet wide, had a sloping lid with two glazed shutters that could be opened to admit air and could be covered with two rolls of tarpaulin in rough weather. This packing case, Lindley claimed, was "the most suitable for importing plants from distant countries" (fig. 2.3).[42]

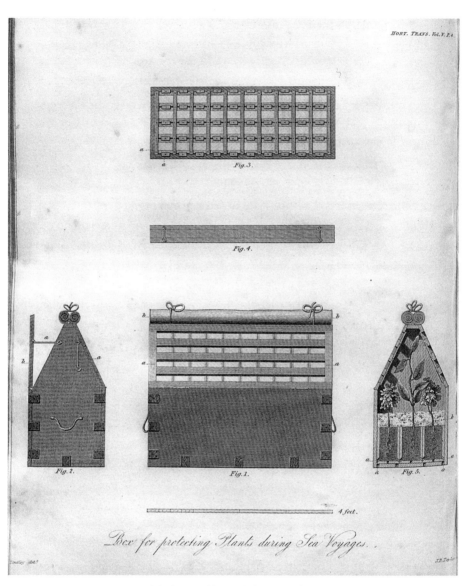

Box for protecting Plants during Sea Voyages.

FIGURE 2.3 Diagrams of the box used by Sir Robert Farquhar to transport plants from Mauritius to London in 1824. This box was promoted by John Lindley at the Royal Horticultural Society, London. From John Lindley, "Instructions for Packing Living Plants in Foreign Countries," *Transactions of the Horticultural Society of London* 5 (1824).

The case that Lindley described was an important development: it was very similar to the Wardian case, the only difference being the shutters. Ward's revolutionary contribution to the plant box was that in his system the case remained closed. Years later Joseph Hooker remarked that before Ward's invention there were other cases similar to his designs, "but these were used rather as sun-shades, and protections against rain and salt water, than as Ward's cases are."[43] Another early variety of plant box was sent to the greenhouses at the Horticultural Society by the pioneering botanist and surgeon for the East India Company Nathaniel Wallich, based in Calcutta. Wallich sent boxes with a roof made with Chinese oystershell inserts to allow the admission of light.

The Long History of the Wardian Case

The challenges of sending live plants across long distances are as old as travel itself. In the seventeenth and eighteenth centuries, despite the many challenges, a variety of plants were transported to Britain and Europe. It was also during this time that a curiosity about the natural world emerged, with such pursuits as natural history and collecting driving a demand for new and useful plants. Even in the late eighteenth century, when travel by sea allowed both people and things greater mobility, moving plants was still a challenge. Many minds focused on inventing a plant box that could successfully move plants around the world. Ellis designed specific boxes for moving useful plants and put them on display in London. Banks commandeered whole ships (or sometimes just the quarterdeck) in his efforts. His grand vision shows how plants were intimately connected to large imperial projects that operated on a global scale—a feature that New Imperialists would adopt a century later. Yet, despite Banks's extravagant ventures, the plant box in its various designs was the most common and useful way to move plants around the world.

Many of the earlier style of cases and boxes are very similar to the ones that Ward used to transport plants. Indeed, much of the advice offered in the eighteenth century was prescient. By charting the longer history of plant transportation between 1660 and 1820, we see that plant boxes were on the move for a very long time. Ward's invention was two centuries in the making and had a long history. But for much of this time there were still great losses when moving plants. The challenge of moving plants did not dampen people's desire for them—quite the reverse: the demand for foreign plants only increased. Even as late as the early nineteenth century, many methods were being tried to overcome the challenges. What Ward showed with his 1829 invention and his 1834 experiment was not only that plants could be enclosed for extended periods of time, but that so enclosed they could better survive long journeys.

3

Global Gardens

In the 1830s George Loddiges championed Ward's invention for moving plants. His nursery, Loddiges & Sons, was the leading European nursery of the day. Following the first successful transplant of plants to Australia and back, he put into use nearly five hundred glazed cases to various parts of the globe. Loddiges's reputation was such that his approval gave Ward the promotion he needed, and it led others to experiment with the cases. "In short, nothing more appears to be wanting to ensure success in the importation of plants," wrote Loddiges to Ward, "than to place [the plants] in these boxes properly moistened, and to allow them the full benefit of light during the voyage."[1]

Throughout the nineteenth century there was a thriving nursery trade in Britain and the rest of the world, and Loddiges & Sons' plant collection was without equal. One commentator noted, "It is probable that no private establishment ever contained anything like an approach to such multitudes of rare species belonging to every cultivable division of the Vegetable Kingdom."[2] Loddiges's experience drew many enthusiasts and professionals to him.[3] A man of deep religious convictions, he would throw his support behind efforts he believed in. He was known for his generosity with scientists in supplying them with exotic plants, as he did for Ward's experiments. He also spent time with young botanists and educated them on the best ways to move plants between countries.

Ward's glazed cases were adopted soon after that first successful movement of plants to Australia and back. It was not directors of botanical gardens or state-sponsored explorers, but nurserymen and private citizens—members of civil society, people like Loddiges and others who led the fashion for exotic plants—that paved the way for the Wardian case to becoming the preferred method of moving plants.

In the 1830s, following Loddiges's adoption of Ward's cases, three important events set it on its path to success. First, the duke of Devonshire, William Cavendish, used Wardian cases to transport plants to his newly constructed greenhouse at Derbyshire. Second, William Hooker, one of the leading botanists of the day, used the cases to transport plants from Brazil to Glasgow, along the way becoming close friends with Nathaniel Ward. Third, the British Association for the Advancement of Science invested heavily in testing and showcasing Ward's invention for their 1836 meeting in Liverpool. It was also at this time that Ward's authority to the invention was challenged by the Scot Allan Maconochie.

Gibson in India, 1836–1837

Gripped by his era's passion for exotics and encouraged by the vision and talent of his head gardener, Joseph Paxton, Cavendish set about building a huge greenhouse. Paxton wanted to build a palm house that was original in design and grand in proportion. For advice they took their designs, including a miniature model, to both George Loddiges at Hackney and later to John Lindley at the Horticultural Society. Confident that it would meet their expectations, they commenced building.

Cavendish and Paxton wanted rare plants to fill their greenhouse that would match the grandeur of the architecture. They selected their young Chatsworth gardener John Gibson to hunt for plants in India. Gibson's primary purpose for the trip was to collect *Amherstia nobilis*, the pride of Burma, a newly discovered tropical tree regarded and sought for its large, extravagant, drooping bloodred blossoms (although they also wanted orchids and other rare ornamentals).[4] *Amherstia* was first found by Nathaniel Wallich in 1826 at a Burmese monastery about twenty-seven miles outside Martaban. Wallich was the superintendent of the East India Company's Botanic Garden in Calcutta and was widely regarded as the authority on Indian botany, and he offered to help Gibson complete his mission.

As Gibson was waiting in London for his ship to leave, Wallich received a consignment of plants from Loddiges & Sons. So impressed was he that he wrote a letter to the *Gardener's Magazine*. But it was not the plants that captivated Wallich; it was the way they were sent to him: in a Wardian case, one of the five hundred Loddiges had put into circulation. Wrote Wallich, "I wish you would make mention of this most extraordinary and novel mode in your Magazine."[5] He would go on to use the Wardian case to transport plants from the Indian subcontinent to his networks in Europe. Significantly, he sent the first cases to France.

Before setting out, Gibson spent time in London with plant specialists to learn about Indian plants. He also made a special trip to Hackney to learn about the glazed cases Loddiges had used and obtained two boxes from him. He filled the cases with valuable plants that were to serve as floral gifts to people he met on the subcontinent. The *Jupiter* sailed for Calcutta in September 1835 with two glazed cases on the poop deck, as well as some open cases of plants. Lacking the confidence of Ward or Loddiges, Gibson worried about the plants sitting in the sun in the glass cases, remarking that he "did not like the appearance of the plants in the so much famed air-tight cases."[6] When he arrived in Calcutta, after nearly five months of travel, the plants in the glazed cases had done well, but those in the open cases had all perished.

In Calcutta Gibson met with Wallich, and soon after arriving they received word from the governor of Bengal that they would also support the collecting trip by giving thirty-two rupees to cover the wages of two local collectors. With this endorsement, the Calcutta gardens would also receive plants from the Gibson trip. Wallich then outlined the journey for Gibson: he would travel by boat northeast to the Khasi Hills in the Assam Mountains.

In July 1836 Gibson set off with two local guides, Ram Chund Maulee and Ramnarain Baugh, traveling by boat up the tributaries of the Brahmaputra River. After arriving at Chhatak on the Surma River, they then went on a monthlong trek on foot to the small village of Cherrapunji. Situated in the eastern part of the Khasi Hills district in northeast India, Cherrapunji is well-known for being one of the wettest places on earth and for its living tree-root bridges that span the rushing rivers. From his base in a small bungalow in town, Gibson collected in the jungle and explored the hills and tributaries of Assam. He wrote to Paxton: "I do assure you that such is the extent and splendour of my collection as to make it one of the richest collections that has ever crossed the Atlantic."[7] It was a gold mine for collecting, with hundreds of varieties of orchids and many other beautiful plants.

Maulee and Baugh made an important contribution to the collection. Not only had they paddled the boat that took them into the mountains, but it was their connections with local residents that allowed Gibson to amass such a large collection of plants. Wallich told the government officials in Bengal who were financing the men on the trip: "The benefit which this garden has derived from the above collectors [Maulee and Baugh] especially in the Orchideous sorts of plants is very considerable." But Wallich was unwilling to credit all the success to the knowledge of the native collectors, giving over to the imperial lenses of the time: "Much of their success is attributable to the peculiar circumstances in which the men were placed under the directions of an English gardener." Nonetheless, the knowledge and efforts of Maulee and Baugh were vital to the success of the trip.[8]

As the Khasi Hills were proving a rich field for collecting, it was decided that Gibson would remain there and that one of the specimens of *Amherstia* in the Calcutta gardens would be supplied to Chatsworth. From the Khasi Hills Gibson sent packages of plants downriver to Calcutta, where they were received by Wallich. These were then packed into glazed cases and dispatched to Chatsworth. Gibson arrived back in Calcutta in February 1837. Maulee and Baugh, who had been so important to Gibson's success, deserted as soon as they came back down from Cherrapunji and never went back to the Calcutta gardens. They never collected the wages owed them for the journey; no details are available as to why.

On 4 March 1837 Gibson departed Calcutta on the *Zenobia* with "12 cases on the poop," six large and six small,[9] and eventually arrived in

England with an enormous collection sourced from the Indian mountains. When he was almost home, he wrote to the duke from Plymouth: "The plants are in the most beautiful order and preservation."[10] Before Gibson's departure Wallich had already shipped thirty glazed cases to Chatsworth full of Gibson's plants. In total the shipment comprised over forty glazed cases—a huge haul for the time. Among the plants were a specimen of *Amherstia* and at least fifty new varieties of orchids, as well as other ornamentals, including varieties of bignonia, impatiens, melastoma, and rhododendron. After Gibson arrived back at Chatsworth he was promoted to foreman of exotic plants. The trip to India was the highlight of his career. It was also the highlight of the early history of Ward's glazed box.

With Ward's cases, many people with a desire for plants were able to optimistically pursue grander collecting expeditions in foreign locations. In doing this a range of interests were drawn together: nurserymen, wealthy noblemen, the British and Bengal governments, merchant vessels, and even local collectors from Assam. They were all connected to the search and transport of new plants.

Following Gibson's success, as well as other successes of Loddiges's nursery, others began to use the case for their plant transports. Wallich saw the success of the glazed cases and commenced sending them on to his colleagues at key scientific centers in continental Europe. Similarly, Lindley, at the Horticultural Society of London, observed the success of the Gibson journey and was now fully convinced of the closed system. In 1836, when the Society sent Karl Theodor Hartweg to California and Mexico, he took with him a glazed case modeled on Ward's design. The case became an important tool for the Calcutta Botanic Garden: between 1836 and 1840 two-thirds of all plants sent from Calcutta were sent using Wardian-style glazed cases. The Wardian case was quickly becoming the most common way to move live plants.

Blossoming Friendships

William Jackson Hooker was one of the leading British botanists of the nineteenth century. He was acquainted with many of the foremost scientists of the early nineteenth century, including Joseph Banks. In the 1810s he got himself into financial difficulty following a failed investment in a brewery and tried to find a way to support his income through his botanical expertise. At the time, paid botanical positions (other than as a gardener) were not only difficult to obtain but also were looked down upon by many in the upper classes. Yet in 1820, through the recommendation of Banks, Hooker secured the botany professorship at Glasgow University. He was a tireless worker and maintained a strong network of correspondents both in the south of Britain and across the world, building up contacts in government, the India Office, and the colonies. To amass his collection and publish his findings, he sourced plants from around the globe.

Like many who came before him, Hooker turned his attention to the problems of transporting plants, even publishing a number of pamphlets on the topic.[11] He first heard about Ward's invention from the botanist William Henry Harvey, a tireless plant hunter who traveled to many colonies in search of botanical treasures. Before one journey to the Cape of Good Hope, Harvey dined with Ward at Wellclose Square. Over a meal on 11 December 1834 they discussed their mutual interest in ferns and mosses, and Harvey was introduced to Ward's experiments with plants under glass. He was so impressed by Ward's invention that the next day he wrote to Hooker telling him about the genial and cheerful doctor from Wellclose Square who was growing ferns in the polluted London air.[12]

Months later in the Cape Colony, Harvey collected a rare moss—there is only one species in the genus—and sent it to Hooker to name. Hooker named the moss *Wardia hygrometrica* and wrote: "I am permitted to join its discoverer [Harvey] in dedicating to N. B. Ward, Esq., an ardent promoter of Botany in all its departments." Hooker went on to describe Ward's major contribution: "[Ward,] who has laid open a new field to the philosophical inquirer, by his method of preserving living plants during long voyages, and of cultivating them in the midst of large cities in closed cases."[13] Such was the naming of nature in the nineteenth century: connections and networks mattered a great deal in who was recognized and remembered (fig. 3.1).

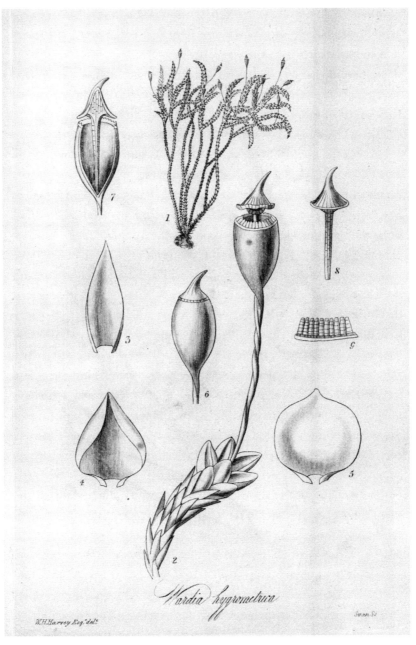

FIGURE 3.1 *Wardia hygrometrica*, 1836. Collected in South Africa by Ward's closest friend, W. H. Harvey, and named by William J. Hooker in honor of Ward. It was through this plant's discovery and naming, and the ensuing correspondence regarding it, that Ward and Hooker became friends. From *Companion to the Botanical Magazine* (1836).

The earliest correspondence between Ward and Hooker was concerned with how to use Ward's case. That first letter, dated 13 January 1836, published in the *Companion to the Botanical Magazine*,[14] was one of the earliest reports by Ward on his system of closed cases, and it received sweeping readership through Hooker's extensive networks. As knowledge of the case spread throughout botanical circles, Ward came into contact with more and more people who promoted it widely.

In 1836 the friendship between Ward and Hooker blossomed. Soon after their correspondence began, Hooker contacted Ward about the best means of using the glazed cases for sending plants from Brazil. One of Hooker's young students, George Gardner, was about to travel to Brazil on a plant-hunting expedition. Ward responded confidently to Hooker: "I believe that in these, ninety five out of one hundred could be imported in a vigorous state from any part of the world, provided the voyage did not exceed eight or ten months in duration."[15] Ward and Hooker decided that instead of taking a glazed box with him, Gardner might do better to have them purpose-built in Rio de Janeiro and only take with him small squares of glass. Ward responded to Hooker's letters openly and technically, passing on his experience: "It may be as well to state [once] and for all that the success of my plan is in exact proportion to the admission of light to all parts of the growing plants and for the due regulation of the humidity of the mould [soil] in which they are planted."[16] As was common in nineteenth-century plant-collecting trips, Gardner's trip was organized as a syndicate. Hooker organized the journey but had numerous individuals, such as the duke of Bedford, subscribe to the trip. The subscribers paid a fee and in return shared in the plants collected upon Gardner's return.

Gardner in Brazil, 1836–1841

For Gardner, Brazil was a fertile field for collecting. He first set out to explore the Serra dos Órgãos, in the vicinity of Rio de Janeiro (fig. 3.2). Over the next four years he expanded his search, in total covering three thousand miles and exploring a large portion of Brazil, including areas in the vicinity of Pernambuco, São Francisco, Aracaty, Ceará, and Piauhy,

FIGURE 3.2 View of the Serra dos Órgãos, Brazil, where George Gardner collected plants for the six Wardian cases that he took back to Glasgow. From George Gardner, *Travels in the Interior of Brazil Principally through the Northern Provinces and the Gold and Diamond Districts during the Years 1836–1841* (London: Reeve, Benham & Reeve, 1849).

making one long journey with sixteen horses, four men, a dog, a monkey, and several parrots.[17] Along the way he compiled a large collection of medicinally and economically valuable plants: some were used to treat smallpox, others (*Aristolochia*) cured snakebite, and still others (*Croton*) treated venereal complaints.[18] His observations of both rainforest and arid regions were keen and practical.

In late 1840 Gardner arrived back in Rio. Staying only a few months, he oversaw the construction of four large boxes, made "somewhat on Mr Ward's plan." He left those boxes with his friend the German botanist Ludwig Riedel who was also the director of the botanical garden in Rio. Riedel had collected throughout Brazil before settling into the botanical-garden position. He and his colleague Bernhard Lushnat collected for the Saint Petersburg Botanical Garden in the early 1830s. On one consignment, they sent almost one thousand "rare and beautiful" live plants to the Russian center of botany. At the time they used Lushnat's innovative method of stripping plants of all their leaves and excess and laying them vertically in clay inside a tin box. The method, although cumbersome, was successful. They were the first tropical plant collections to arrive in the Saint Petersburg Botanical Garden in prerevolutionary Russia.[19]

Gardner spent March and April 1841 collecting in the thick forests and in the high mountains. One of the plants he collected was "a great beauty," a scarlet-and-white-flowered shrub that Gardner was anxious to bear Hooker's name (it went on to be called *Prepusa hookeriana*). Once he had finished collecting in the Serra dos Órgãos, he headed back to Rio to pack up his collection of living plants. He had so many plants that he needed two more Wardian cases made. On 5 May 1841 Gardner had his six cases installed on the deck of the slave ship *Gipsey*.[20]

The Wardian cases sat on the ship deck gathering sunlight while men, women, and children were chained below deck. After fifteen days at sea they stopped at Maranhão, a state in northeastern Brazil. Despite their privileged position on the deck, some of the plants were not doing well. Here they waited for three weeks, presumably swapping slaves for cotton.[21]

Gardner arrived in Liverpool with six large Wardian cases, and even though it had been only a thirty-two-day journey, many of the plants in the cases perished. In Liverpool it was "awfully cold." Gardner wrote to Hooker that the plants in the cases "have not done so well as I could have expected—out of 150 species, not more than 50 are now alive. . . . Many fine things have perished, but there still remains much that Mr Murray [the gardener at Glasgow] will be proud of." The beautiful *Prepusa* that had been named after Hooker was not doing so well, but he hoped it would make it to Glasgow alive.[22]

Gardner's journey was a partial success. In five years he collected nearly sixty thousand herbarium sheets consisting of nearly three thousand species, which he passed on to subscribers or sold. Although the herbarium specimens arrived in good condition, the loss of so many live plants was a sore point for the botanical gardens in Glasgow. The failure stemmed from Gardner's haste in collecting plants too soon before sailing and the hot, humid weather in Maranhão.[23] Much hype surrounded Ward's invention, but Gardner's shipments show that although there were many successes, there were also failures.[24]

The five years, 1836–41, that Gardner was in Brazil was a long time in the biography of the Wardian case. It was a time not only of further experimentation and proclamations in support of Ward, but also of mounting challenges to Ward's scientific authority.

On Show

Ward was not a prodigious publisher, but he was a great networker and used his connections to promote his invention. In 1836 he sent a letter to Reverend James Yates, the secretary of the newly formed British Association for the Advancement of Science, telling him about the glazed cases. Following a discussion among members of the British Association, £25 was set aside so that a glass case modeled on Ward's plan could be constructed to greet members as they arrived at the 1837 Liverpool meeting of the British Association.

This was quite an honor for Ward. The Association had been formed in 1831 with the aim of giving systematic guidance to scientific research and to lobby government on its advancement. Yates wrote to his fellow organizers at the British Association about the possibility of Ward's cases being on show: "They would be a novel, attractive and beautiful object of attention, and at the same time be an aid to botanical science on account of the facilities which this method affords for bringing plants from abroad." In the lead-up to the Liverpool meeting, a committee made up of Yates, the Oxford botanist Charles Daubeny, the Cambridge professor John Henslow, and the inventor from Dublin University Museum Robert Ball formed to oversee the case project. Ward and Hooker were advising.[25]

In September 1837, in preparation for the meeting of the Association, Yates constructed a glass plant box nine by eighteen feet according to designs delivered from Ward. Positioned at Mount Street, in the courtyard of the Mechanics' Institute, and filled with eighty species of plants on loan from the Liverpool Botanic Garden, it was a centerpiece of scientific experimentation on show for scientists from all over Britain and the Continent. Yates knew the results would be favorable: he had grown plants under glass in London in the lead-up to the event. The Association had grown in popularity since its first meeting. Opening day in Liverpool attracted over five thousand visitors, and the opening lecture, delivered on the Monday evening, had almost three thousand attendees. Over the course of the meeting Ward's invention was seen by thousands of people, all interested in the latest scientific discoveries.

The committee also reported on their various scientific experiments related to the case. Ward delivered an address in which he talked at length about his discoveries and experiments. He also used the occasion to clarify a growing misconception about sealing the cases: "In no instance have I ever endeavoured to seal the cases hermetically; it would, I conceive, be almost impossible to do it, and if done, would prevent that continued change of air from its alternate expansion and contraction, upon which in my opinion the success of the plan mainly depends."[26] Above all it was the presentation of the actual case flourishing and resplendent, on show to many of Britain's leading scientists, aristocrats, and politicians, which gave it visibility and convinced them of the possibilities.

With the favorable reception of Ward's concept for closed cases, it seemed smooth sailing toward their technological uptake. But as more and more people wanted to use the case, many people wanted a piece of Ward's time. Often people wanted specific details on how to construct their own case or even just to look at the different plants Ward kept in his cases. One keen gardener was James McNab, the superintendent of the Royal Caledonian Horticultural Society Garden in Edinburgh, who visited Ward numerous times so that he could perfect the design and show off the new cases to Scottish gardeners and botanists.

On 13 June 1839, at the meeting of the Edinburgh Botanical Society, a glazed case based on the designs prescribed by Ward was shown. It had been built by McNab for the botanist Daniel Ellis, who delivered a long lecture on the topic of growing plants in glass cases. Most compelling about the presentation of this case in Edinburgh were the specific details of its construction, which were provided from McNab's close contact with Ward. The edges of the box that contained the soil were made of well-seasoned Santo Domingo mahogany. The bottom of the soil box was made of inch-thick Honduras mahogany; to add extra strength, two cross-pieces were inserted across the width of the box. The bottom of the box had a number of small holes in the bottom to allow for drainage. A groove was sunk along the upper edge of the box so that the glass roof could be inserted on top. Inside the box McNab first covered the bottom with about two inches of broken pots and rocks, on top of this was an inch of fibrous and thick loam, and then the remaining depth of the box was filled with soil, watered with four gallons of water, and then allowed to drain for a day. The box was then planted with a number of European, North American, Asian, African and South American plants, including mayflowers, pitcher plants, and rhododendrons.[27]

Following the presentation of the case, the Edinburgh Botanical Society congratulated Ward for his "zeal and perseverance." The Society concluded that the botanical enthusiast had now "acquired the means of transporting to their own soil the varied and most delicate plants of every region of the earth." At each turn, Ward's invention received wide praise from scientific circles.

Ward's Case

It was with some surprise that, following Ellis's paper, McNab, who had himself spent much time with Ward, stood up to read a letter written to the society containing a challenge to Ward's authority over the system of closely glazed cases. In his letter to the Edinburgh Botanical Society, Allan Maconochie said that he had practiced "window gardening" in glass cases for at least the last fourteen years with much success. He was inspired to experiment with plants after hearing of the experiments by de Saussure and also of Humboldt's discoveries of plants deep in the ocean. He started by using a large glass vessel, in which he enclosed a number of foreign ferns. Following this success, he had a carpenter make him a "miniature greenhouse." His friends had all seen and witnessed his method of growing plants. Maconochie wrote that he "had no wish to detract from Mr Ward the merit of also having made the discovery, or that of having made it known to the public." It seemed that Maconochie was magnanimous in referring to Ward as having claim to the case, but he did want it on record that he had already been using such a system well before Ward.[28]

These were challenging times for Nathaniel Ward. It must be remembered that Ward was only an amateur naturalist, although his invention was garnering wide coverage; throughout Ward continued to work at his busy medical practice in Wellclose Square. On 15 October 1839, a few months after Maconochie's letter was read to the Society but well before Ward heard about it, a three-month-old baby named William Watts was brought to Ward's practice for a routine vaccination. A few days later the mother became concerned about her child; his arm was inflamed, his chest was swollen, and the infection was spreading to his other arm. Believing the injection had not been given properly, she took the baby back to the doctor. On Sunday, 3 November, young William died in agony. The mother believed the child's death was the result of Ward's malpractice, and the following week he was deposed. Ward had been a government vaccinator for twenty years; he had vaccinated about 24,000 children, and this was the fourth death. Ward told the court that he was quite frankly "surprise[d] that many more children did not die after vaccination, as proper care was not taken in nursing them." The court deemed it a "natural death," most likely from a bacterial infection following the vaccine; Ward was free to return to his practice and take up botanizing in his spare time. But soon after young William's death Ward himself contracted an infection caused by an open wound on his thumb from which it took him three months to recover.[29]

Given Ward's challenging circumstances, news of the Maconochie letter, which filtered through to him in the early months of 1840, was quite shocking. While Maconochie acknowledged Ward's promotion of the glass case, the normally considerate Ward was deeply offended by the publication of Maconochie's letter in the *Proceedings of the Botanical Society of Edinburgh*, interpreting it as a claim that botanists in England had taken Maconochie's plan and were promoting it as their own. It was this point that brought out the worst in Ward.

Writing to Hooker, Ward said that the Maconochie letter was "one of the most barefaced falsehoods ever uttered, and Mr M. himself might

well know to be such." He accepted that Maconochie might have experimented with the case for longer than he had but refused to accept that
his discoveries were not original. In his defense he listed many of the
experiments that had taken place, from the journey to Sydney to the
showing at the British Association meeting. He also noted the many
people he showed through his house to see the cases and pointed out
that none of them had ever mentioned the cases in Edinburgh. Maconochie had never, Ward continued, "stepped forward to assert his claim
of priority, although he has had ample opportunities of so doing, as the
utmost publicity has been given to my publications." His tone turned
from defense to despair when he noted near the end of the letter that "I
had never rec.d or wished to rec. from anyone the slightest recompense."
In the end, Ward claimed that the only prize was the endless reception of
visitors or answering inquiries regarding the case.[30]

Feeling affronted and abandoned, Ward came out of the Maconochie
dispute energized to lay claim to his invention. Ultimately he received
encouragement from Hooker and his other friends. At Ward's insistence,
many of his contacts in scientific circles credited him with the box's
invention. Maconochie is largely forgotten in the history of the glazed
plant box; after his brief letter to the Edinburgh Botanical Society, he
published only one further account related to gardening under glass. The
Maconochie affair, however, shows that while Ward was often magnanimous and deferential to the politics of scientific matters, on the subject
of the glazed case he wanted to remain closely attached and remembered.

Following this incident Ward set about completing his only monograph, *On the Growth of Plants in Closely Glazed Cases* (1842), which he
had been putting off for over a decade. He also set to using his wide networks in the promotion of his invention: as he wrote to Hooker, "If it be
not giving you too much trouble will you be kind enough to furnish me
with a list of such of your foreign correspondents . . . as may be interested
in my experiment. As I wish to send a copy of my little work to each of
them."[31]

To be sure, Ward came up with his ideas in isolation; it often happens
that inventions are conceived concurrently at different places by different
people. Ward deserves to be credited as the box's inventor. Not only did

he come up with the idea, but it was his publications and promotions of the case that ultimately cemented its place as a key botanical technology. By sending out his book to many foreign correspondents—just one of many examples of his promotions—he ensured that this method of transporting live plants would forever memorialize him as the inventor of the boxes. It was after the Maconochie incident that Ward's name, particularly in London scientific circles, became closely attached to the case, and the name underwent several changes, from "closed glazed cases" to "Ward's glass cases" to simply "the Wardian case."[32] It was in these years between 1839 and 1842 that the name "Wardian case" came into common usage.

Scrutiny of Kew

Ward's friend Hooker was above all most important to his association with the glazed case. Their friendship blossomed out of correspondence concerning the Gardner expedition. In Glasgow, not only was Hooker a professor of botany, but he also oversaw the botanical gardens, wrote popular and scientific books, edited many magazines, and sourced plants from many parts of the globe through a wide network of collectors. Hooker is most remembered for his directorship at Kew Gardens, but it is often forgotten that before his appointment at Kew he spent two decades forming connections and administering a successful botanical empire from the north.

In 1838 Kew Gardens were in a state of disrepair. Decades of neglect since the death of Joseph Banks had left the eighteen-acre garden and the attached two-hundred-acre game reserve as a site of concern and potential redevelopment. Appointed to assess the value and future of Kew Gardens were John Lindley, Joseph Paxton, and John Wilson, gardener to the earl of Surrey. They toured Kew in February 1838 and they found a garden that merely raised seeds and looked after plants. Although claiming to be the great botanical garden of Britain, it did not conduct scientific research or teaching, it did not communicate with the colonies, and only recently had it even engaged in naming the many plants that were sent to it.[33]

Although the records before 1841 were poorly kept, a fact Lindley observed in his report on Kew, a few consignments of plants left Kew in the first three decades of the nineteenth century. From 1806 to 1838 a total of 483 deliveries of plants were sent, of which only 28 were sent to the colonies; a further 171 smaller consignments of seeds were sent to foreign plant collectors. From 1830 to 1838 Kew sent out only two packets of live plants to the colonies. William Aiton, the head gardener, who was under fire, claimed that "the Garden cannot be saddled with the expense of fitting up boxes for exportation."[34]

Aiton, who became the head gardener of Kew following his father's death in 1793, was keenly aware of the committee's harsh appraisal of his management. Knowing that his lack of contact with the colonies was a particular problem, it is no coincidence that on 1 May 1838, only a few months after the committee's visit, the first Wardian cases were sent from Kew. Four glazed cases, following Ward's method, were sent to the "new colony" of Western Australia. They contained eighty-four plants made up of everything from apple to cherry, gooseberry to quince, balsam to red cedar, and ash to oak.[35]

Sending out belated consignments of valuable European plants to Western Australia was not enough to save Aiton. Things had changed in the four decades since he had become director: the science of botany was much more advanced and systematized, and gardens were now in service of many interests, including the state, the economy, and the public.

After his visit to Kew, Lindley envisaged a national botanical garden that would be at the center of a British botanical empire. Already there were gardens in Bombay, Calcutta, Sydney, Saharanpur, and Trinidad that all demanded plants; added to this, colonists in many locations were also in need of plants and information. All the far-reaching colonial gardens would work within this system. A national garden would be able to be in the service of the government, providing information on the environmental conditions of potential new colonies and supplying plants to such colonies; it would also be a place to process the new and valuable plants from abroad that could then be dispersed throughout Britain.

Following the committee's tour in 1838, the Kew agenda was firmly on the minds of government. After much indecision from politicians, a renewed Kew emerged. What it most needed was a new director. The decision came down to a choice between Lindley and Hooker.[36]

In April 1841, at the age of fifty-seven, having the support of the duke of Bedford and offering to take a lower salary than Lindley, Hooker assumed the directorship at Kew. He oversaw the transport of his vast library and herbarium from Glasgow to London, and he also took his many networks and contacts with him. Only a few months after commencing his role as director, on 9 June 1841, thirty-three plants were sent from Kew to Ward at Wellclose Square. Following Hooker's arrival in London, his contact and friendship with Ward grew, and so too did the use of Wardian cases for sending plants.

Aiton sent out the first glazed cases in 1838 and 1840, but so far none had been received at Kew. In April 1842, only a year after Hooker took over, Kew received a "glazed box" from William Symonds, a deputy surveyor general in New Zealand. Symonds's father, a well-known naval captain and surveyor of the same name, was an acquaintance of Hooker's and had forwarded the box to Kew on behalf of his son. This was quite significant, for not only had the plants arrived "fresh," but they were planted out in the orangery, including a specimen of the rimu tree, *Dacrydium cupressinum*. By 1857 this slow-growing conifer, which flowers with tiny, beautiful red bursts, was thriving in the orangery, having reached a height of eighteen feet.[37] Hooker's move to Kew was an important moment in the history of the Wardian case. Hooker was a friend of Ward, he had firsthand experience with the cases, and in the 1840s he began promoting them in his official publications to the Admiralty. Over the following years, Kew's use of the cases increased dramatically.

From the outset we can see that it was not just physical plants that were being moved; networks of people were forming that allowed for greater numbers of plants to be sent. Ward was protective of his reputation because of his friendships with many scientists, friendships that developed not because he was a prodigious publisher, but because he was a caring friend and excellent networker. In just a decade following its invention the Wardian case had traveled the world and been instrumental in moving many plants. It started at Loddiges's nursery at Hackney and spread through a wide circle of the most influential botanists and horticulturalists in Britain. While Loddiges's approval was instrumental in getting the case recognized as a prime mover, it would be Hooker and his role as the director at Kew who would take the case to the colonies and show its utility. From the outset the Wardian case required the emerging networks of trade, science, and colonization to reach around the globe. As these networks increased in reliability, reach, and frequency, so too did the use of the Wardian case.

4

·······················

Science at Sea

On a cold and frosty Saturday in early February 1839, Ward sought the warmth of Loddiges's greenhouse. He met three young botanists in London, and together they took the omnibus to visit the wonderful greenhouses at Hackney. It was a "rather long" journey from the city to the northeastern parish. In the carriage with Ward were the twenty-two-year-old Joseph Hooker; his younger sister, Maria Hooker; and the nineteen-year-old American Asa Gray. The Hookers were the children of William Jackson Hooker and shared their father's passion for plants. Joseph postponed his return to Glasgow to make the visit to Hackney; Maria had already traveled from outside the city just for the excursion; and Gray was in London visiting the best British collections in order to compile the expanded second edition of his *Flora of North America*. Ward was kind and encouraging to the young botanists and took them to see his friend Loddiges.[1]

Gray and Joseph Hooker would go on to become two of the most important botanists of the nineteenth century. While Maria was not destined to pursue life as a professional botanist, she was a keen amateur who continued her excellent education throughout her life. Later in the same year in which he met Ward and Loddiges, Hooker was appointed to the British Ross expedition, which was preparing to leave on a four-year journey of scientific exploration in search of Antarctica and the magnetic South Pole. The Americans were also in search of Antarctica, and Gray was appointed botanist to the United States Exploring Expedition, often simply called the Wilkes expedition after its leader, Charles Wilkes. However, he chose not to go on the expedition and accepted a professorship in botany instead.

Loddiges was very accommodating to the visitors and spent the day showing them the Hackney nursery's many treasures. Ward, Gray, and the Hookers saw the large and beautiful collection of orchids and visited the magnificent palm house. After showing them the greenhouses, Loddiges took them to his house, where he showed them his world-renowned collection of hummingbirds. In his journal Gray wrote of the hummingbirds, "You can't imagine how beautiful they are! They are his great pets." After spending the day at Hackney, they returned to the city, and that evening Joseph Hooker was on a train back to Glasgow.

The day at the Hackney nursery was a propitious moment, at which two generations of plantsmen walked through its grand greenhouses. It was also a time when botanical enterprises were shifting from amateur to professional; Gray and Hooker would both go on to be the leading botanists of their day.

Equally important was the impression the Wardian case made on the younger botanists. At Hackney, Gray and Hooker saw the global effects of using Ward's cases for moving plants. They had both witnessed the case in use in New York and Glasgow, but at Hackney they saw how it was being used by nurseries to amass grand collections.

Shortly after his arrival in London in January 1839, Gray met Ward and saw his boxes. Early in their friendship Gray called Ward "the plant case man" (and "one of the most obliging men I ever knew"), noting that his invention "attracted much attention."[2] A few nights before the Hackney visit, on 29 January 1839, Gray dined at Wellclose Square, where the two men spent much time examining Ward's glazed cases. It was the middle of the London winter, and as Gray observed, "[Ward's] house, which is in the heart of the city, near London Docks, is very badly situated." Gray learned a lot from Ward about packing plants for transport. They would go on to correspond for the next twenty-five years.

The moment at Hackney also made an impact on Hooker. Soon after the Hackney visit he left on the Ross expedition. The Wardian case was the key means for him to send exotic plants to his father at Kew. While at sea he and Ward corresponded. In one letter, dated more than two years after the Hackney visit, just after Hooker had returned from his second descent into the Antarctic ice, he recalled the day they spent together at the nursery: "Remember me to all Botanists who may take an interest in me and particularly to Mr. Loddiges, whose attention to my sister and myself conjointly with your own, I have not forgotten."[3]

Sailing between 1838 and 1843, the first government-appointed expeditions that used Wardian cases were the French d'Urville expedition, the British Ross expedition, and the American Wilkes expedition. The French and Americans had acquired their knowledge of Wardian cases through Loddiges's nursery. Hooker traveled as the botanist on the Ross expedition, and Gray wrote up the collections of the Wilkes expedition and remained forever linked to its botanical results. Before setting sail with these expeditions, let's first consider the sea.

Boxes at Sea

Saltwater was a curse for live plants on a long sea voyage. One of the great assets of Ward's cases was that they made it unnecessary to leave open cases on the ship's deck. The enclosed case allowed plants to sit comfortably on the poop deck, with the glass both admitting light and repelling sea spray. The enclosed system also required little attention from sailors and did not use precious freshwater resources. While it was a major step forward for moving plants across oceans, it was not, however, as if the case looked after itself.

Often the most successful transfers of plants were those that were accompanied by a concerned ship's captain or gardener. Take the first case ever, sent in 1833, that went on the *Persian* with Captain Mallard. In his second letter to Ward from Sydney congratulating him on the success of the cases and telling him of the transplant of the British plants to the Sydney gardens, Mallard wrote of the "pride and pleasure" that he felt at having been "the instrument" to prove the importance of this discovery to the botanical world and concluded, "PS. I ought to have mentioned that during the voyage the plants were watered but once, and that but a slight sprinkling near the Equator." Although Ward included the letters from Mallard as key support for the success of his invention for transoceanic plant movement, he neglected to include this postscript either in his full-length monograph in 1842 or when it was reissued in 1852. Those caring for cases on a ship played an important role.[4]

Loddiges praised Mallard's work in the transport of plants: "Amongst all we have sent out or received, none have arrived in such good order as those brought by this gentleman [Mallard]. I wish we had more that possessed his love for Natural History, and would take the same care which he has done." Many plants were destroyed because they were not cared for properly once on the ship. Often captains gave Loddiges all assurances, even getting the nurseryman's approval as he watched them set sail with the cases safely on the deck. But once at sea the cases were often put below deck. Loddiges also found from experience that "there cannot be a worse mode of sending plants, than in these same cases, so placed in the dark." Ships and seamen working them were critical to the successful movement of plants.[5]

The importance of ship captains in plant transport was not lost on Nathaniel Wallich at the Calcutta Botanic Garden. When sending plants, Wallich warned, people should not send cases that were too large or heavy, and they shouldn't crowd the cases with too many plants. The cases should also have a neat appearance, "as Captains are very unwilling to allow the deck to be occupied by unsightly objects." Wallich also rewarded captains for successfully moving his plants. Often he went so far as to give a captain one box of plants for every three or four that successfully arrived in London. While many thought that the Wardian case had solved the problems of transporting plants, it was far more complex than a simple matter of case closed. Transporting plants was a practice in which botanists and seaman collaborated.[6]

In the late 1830s there was a renewed interest in scientific journeys of exploration, and the Wardian case became an important new instrument to have on deck. Sailing between 1837 and 1842, d'Urville, Wilkes, and Ross expeditions were all in search of the magnetic South Pole, and with any luck they would discover and map the great southern continent. Each expedition also had national interests accompanying their plans. The French wanted information on the islands in the Pacific; the British wanted to set up a series of observatories for magnetic observation in the Southern Hemisphere, and the Americans wanted a thorough knowledge of the west coast of their continent, particularly the area that would become the Pacific northwestern states of Oregon and Washington. The French and Americans were also protecting their commercial interests in the Pacific. Geographical ambitions were fine, but one of the most important results of all three expeditions was the botanical treasures that they brought home. While many historians have discussed these three journeys, particularly their descent into the Antarctic ice, few have recognized the use of the Wardian case on board their ships.[7]

Serres de voyage: The French d'Urville Expedition

The first Wardian cases to arrive in France were modeled on those of Loddiges's nursery. Sent from Calcutta by Wallich, they were eight months at sea before arriving in Paris in 1836. The cases were sent to the director of the Jardin des Plantes, Charles François Brisseau de Mirbel. The garden, located on the Left Bank of the Seine in the fifth arrondissement, was France's premier garden and its center of botanical and horticultural research. In Paris Mirbel excitedly opened the never-before-seen glass cases and found inside fifteen species of plants that "scarcely appeared to be more exhausted than plants taken out of the green-house at the return of spring" (fig. 4.1).[8]

So impressed were the French botanists that they copied the design of the cases. Like all plantsmen respecting the unwritten laws of botanical exchange, they returned Wallich's case filled with plants from their collection. Mirbel also sent news of the case to others, among them the famous Cel nursery family, well-known for importing exotics to France. The Cels also sent Wallich a Wardian case as thanks for introducing them to the new case. Indirectly, the knowledge of the Wardian case came to the French from Loddiges's nursery: Wallich had learned of the cases when he was sent plants from Loddiges the previous year, and he then

FIGURE 4.1 Wardian cases were used by the French for over a century. This case was preparing to leave the greenhouses of the Jardin d'Essai Colonial, Paris, in the late nineteenth century. From *La dépêche coloniale*, 15 August 1903.

sent cases to Paris. When Mirbel described the new Wardian cases he was quick to thank not only Wallich but also Loddiges, "who possess[es] the richest nursery garden in Europe."

A leading plant physiologist in his own right, Mirbel was also a botanical advisor for the Académie des Sciences. It was the Academy, an institution closely connected to the state-funded expeditions, that was responsible for issuing the instructions for scientific methodologies to be followed. In the summer of 1837, as the French expedition was preparing to leave Toulon, the Academy advised on the expedition's scientific objectives. Although some leading French scientists thought the expedition ill equipped and too hurried to achieve adequate scientific outcomes in the polar regions, the Academy still gave instructions on astronomy, botany, geology, physics, and zoology. The botanical instructions were by far the most extensive. Mirbel noted that one of the best ways to observe a valuable plant product was to watch what local people were doing: "Wherever man is seen labouring to obtain from the earth, crops adapted to his wants, the form of the instruments of husbandry, the agriculture, the plants which are cultivated, and the produce obtained should be the objects of careful examination."[9] The instructions highlight the intelligence-gathering purposes of the expedition and reveal a keen interest in acquiring indigenous techniques and products.

Mirbel also offered clear instructions on how to transport plants, and he believed the best way to achieve this was in Ward's cases. Mirbel labeled Ward's cases *serres de voyage*—traveling greenhouses. He described in great detail the early designs of Ward's cases, which had a wooden base and a glass structure over the top. For Mirbel, the cases needed to be small: they were to be less than a meter in length and not reach above the height of the hip of an average-sized sailor. The cases could come in different sizes, but "they may not encumber the sailors in the working of the ship which might indirectly endanger the existence of the plants." The cases could be handled by one strong sailor. Mirbel also recognized that differing conditions inside the case, particularly with regard to soil, were needed when sending different groups of plants. Soon after Mirbel penned his instructions the Academy dispatched Wardian cases to Toulon to be carried on the expedition ships, the *Astrolabe* and the *Zélée*.

Helming the French expedition was Jules-Sébastien-César Dumont d'Urville, often considered France's James Cook. He was the leading explorer of his day, a skilled navigator and leader who began his career as a botanist. One of the earliest nineteenth-century French expeditions was d'Urville's first voyage to the Pacific, on the *Coquille*, undertaken for the express purpose of finding new lands or islands where "French ships could transplant civilization and its benefits," as the newspaper *Le moniteur universel* put it. D'Urville would return with the French message two more times, culminating in the journey of 1837–40. The ships sailed for the Pacific on 7 September 1837.[10]

D'Urville first took the two ships to Antarctica after rounding Cape Horn. After months spent battling ice and freezing conditions, the ships descended no further south than the British sealer Weddell had a decade earlier. They spent the following year exploring Chile and several of the Pacific islands. They then sailed to Guam and the Moluccas before arriving in Australia to prepare for another attempt at the ice. Realizing that they were too late to travel south, d'Urville decided to map significant navigational points and coastlines in the Indies before returning to Hobart via the Indian Ocean. On their second attempt south they had more success, reaching 66 degrees latitude and sighting land for the first time. With an eye toward French colonial ambitions, they traveled to New Zealand and mapped the east coast. The ships then sailed home via the Torres Strait. Having spent three years at sea, they arrived back in the harbor of Toulon with a substantial scientific collection.

Among the collection were zoological specimens and plaster busts of indigenous people that captured the attention of French politicians. There were many live plants, which were transported to the greenhouses of the Jardin des Plantes. The minister of the navy, Admiral Guy-Victor Duperre, visited the collections on 28 June 1841 and was impressed. As one observer in the *Annales maritimes et colonials* wrote, the botanical collection "gathered examples of the vegetation of all the points that they had visited."[11] At the time of d'Urville's unfortunate death in a train accident in 1842, he had only seen two volumes of the report on the expedition through publication. Writing up the botanical section of the journey was left to other members of the scientific team.[12]

For the story of the Wardian case, the most important point here is that the French even had the cases on the expedition. They arrived on the French expedition through the informal channels of Wallich, who sent examples of the cases to Paris in the years before the expedition sailed. The knowledge of Ward's cases came from the Loddiges' London nursery. The French would continue to use the Wardian case for many decades.

Joseph Hooker and the Ross Expedition

Joseph Dalton Hooker knew how to pack a plant box. Few could have been better prepared to deploy the Wardian case. He had worked with them in his father's Glasgow botanical enterprise, had examined them firsthand at Ward's house, and had seen them when he toured Loddiges's nursery. Despite his experience, even Hooker at times found Wardian cases difficult to send.[13]

Captain James Clark Ross was an exceptional scientist. On an earlier expedition he had served as botanist; this time he was the leader. Ross was occupied with magnetic investigations and left most of the botanical investigations to Hooker. The Royal Society was the leading scientific institution in Britain, and just like the French Academy, they offered instructions. For the botanical department they recommended collecting seeds. But many seeds, like conifers, suffered when transported, so young plants should be sent home in "Mr Ward's glazed cases."[14] Although the Society advised Ross and Hooker to take only one case, both ships of the expedition, the *Erebus* and the *Terror*, were equipped with two large, cumbersome cases.

Ward was not pleased when he learned of the Society's instructions and complained to Hooker: "I felt at the recommendation of the R. Soc. as to the one case only when there ought to have been fifty or at least the materials for 50 on board all of which might I am certain have been employed to the greatest possible advantage."[15] He went on, "Some [of] these should have been used for the purpose of conveying to our distant colonies numbers of useful timber trees, or other useful plants which have not yet been introduced." Nonetheless, he used his influence on young Hooker to encourage him to use the cases and commended him to pass the information on to Captain Ross.

The expedition left in October 1839. One of their first ports of call was St. Helena, a small island that acted as a stopping place on the journey around Africa's Cape Agulhas. Hooker's botanizing was relatively unsuccessful, but he managed to make new friends. One was a Mr. Wilde, who asked if Hooker could arrange for seeds and roots of common English trees to be sent to him. If these could be sent, Wilde would return the Wardian cases filled with unique trees and plants from the island. On the Ross expedition Hooker was also an agent for his father's expanding networks. At the many stopping places, where Ross was busy building magnetic and astronomical observatories, Hooker not only collected plants but also collected amateur botanists who could send plants to Britain.[16]

In May 1840 the expedition arrived at Kerguelen Island, a small archipelago between Africa and Australia in the southern Indian Ocean. By the time they left the windswept island, the large Wardian case was full of new species of plants, including the Kerguelen cabbage (*Pringlea antiscorbutica*), a unique plant native to the island that was often cooked and eaten by sailors to fight scurvy. At sea there was bad weather. Recog-

FIGURE 4.2 The Ross expedition in the pack ice during a gale, 1840. From James Clark
Ross, *A Voyage of Discovery and Research in the Southern and Antarctic Regions,
during the Years 1839–1843* (London: John Murray, 1847).

nizing the importance of the botanical collection, after a "long consult"
Ross thought it best to put the Wardian case in the great cabin, close to
the windows, where it would still receive light but have greater protection from the elements. From Kerguelen they sailed for Hobart, at the
southern tip of Australia.

The case remained in the captain's cabin until only a few days short of
making land when a fine day gave them relief from the rough seas. Ross
ordered the Wardian case moved back up to the deck. Shortly after the
difficult task of moving the heavy case, a huge storm suddenly descended
(fig. 4.2). They were driven three days off course. Even worse, the plants
suffered terribly: "The late gale came on so suddenly that to have removed the hatches [to put the case below deck] was utterly impossible

the consequence was that one of the seas that broke into us also broke into the case & drenched them all."[17] Only five plants from Kerguelen Island survived; fortunately, the cabbage was among them. When they finally arrived in Hobart, Hooker planted out the surviving plants and hoped they would recover.

Wardian cases were difficult to manage on the deck of a working ship. As Hooker described in a letter to his father from Hobart: "Wardian cases are sadly ticklish things to take to sea, our hatches are so small that they cannot be taken below when full, they are a sad annoyance to the first Lieutenant, who is luckily a good friend of mine, a good sea breaking over us demolishes the glass at once." Maybe the problem was the size of the cases that the Ross expedition took with them; they were much larger than many of those Loddiges used. Still, Hooker's reflections on the cases are insightful: heavy, difficult to manage given the workings of the ship, and in a wild storm susceptible to the elements.[18]

Hobart was a fertile stop for Hooker. He spent months botanizing with a local resident and amateur naturalist, Ronald Campbell Gunn. This connection lasted a lifetime and became instrumental in Hooker's work on the flora of Tasmania. Among their collections were specimens of algae that Hooker secretly posted to Ward at Wellclose Square. Hooker was permitted to send collections only to his father and the government, so sending them to Ward showed Hooker's deep affection. In return, Ward sent many long letters, some containing detailed instructions on packing the plant cases and others informing Hooker of the gossip in London's natural history circles.[19]

The Ross expedition went to Antarctica three times, leaving from Tasmania, New Zealand, and the Falkland Islands. These were not times to find plants. At one point, on the second journey south, they were stuck in the pack ice for forty-seven days, which gave Hooker plenty of time to analyze the plants he had collected in New Zealand. Ultimately the Ross expedition descended farther toward Antarctica than any before them. They reached 78 degrees south latitude, a feat not bettered for another fifty-eight years.

After their first descent south they went to New Zealand. It was around this time that Hooker heard of his father's official appointment as the director of Kew Gardens. He wrote in late 1841: "Most sincerely do I rejoice with you on your Kew appointment." It was not the season for transplanting, but Hooker could not resist sending his father a welcome gift from New Zealand. He asked Captain Ross for special permission to send some plants to his father. Ross assented "at once" and also "gave me hands to remove and fill the box." They sent the large Wardian case from the *Erebus*, and together Hooker and the ship hands filled the box mainly with ferns and other plants from New Zealand: "These were all gathered and packed with my own hands and if they all do as well as they look . . . I shall be well content." The case left aboard the *Exporter* on the morning of 23 November 1841. It went to Sydney and was then sent to London. There is no record of its ever arriving at Kew.[20]

After the second trip into the ice Hooker had more luck. The expeditions stopped at Cape Town and then moved into the Falkland Islands. Hooker collected at both places. He was impressed by the grasses on the Falklands, which he believed had the potential to become fodder grasses, and collected seeds and plants of many species. Among them was the tussock grass (*Dactylis caspitosa*, today *Poa flabellata*), a favorite of imported cattle. Following Hooker's return, tussock grass became an important fodder, widely used throughout the empire. In the Falklands Hooker packed two Wardian cases full of plants. There were ferns (*Lomaria* spp.) and winter's bark (*Drimys winteri*), a species native to Chile often used for furniture and instrument making. There were also "several other curious plants," among them many varieties of grasses, including the strongly scented sweetgrass (*Hierochloe redolens*). Fortunately, these plants survived the journey to London and arrived at Kew in March 1843. They were the third consignment of Wardian cases ever received at Kew.[21]

Hooker's attempts to expand his networks in the southern oceans continued throughout the four-year journey. In St. Helena there was Mr. Wilde, in Tasmania there was Ronald Gunn, and in New Zealand there was William Colenso. And even at the end of the journey, when the expedition stopped at Cape Town, he spent time networking with botanists, first with Carl Ferdinand Heinrich, Baron von Ludwig, who started the first botanical garden in Cape Town, and then with Joseph Upjohn, a nurseryman.[22] For both of these contacts, an initial Wardian case from Britain was what they hoped for to commence ongoing exchanges.[23]

The Ross expedition arrived home after four and a half years. They had compiled much data on magnetism and oceanography and collected many botanical and natural history specimens. Along the way Joseph Hooker also became connected with many people who would become important suppliers of plants and who served the Kew empire over the coming decades. Many Wardian cases were used on the expedition. Some went missing across the oceans, others got left on the docks at Woolwich and did not survive, and others successfully landed at Kew. The Wardian case, as used by Joseph Hooker, was a "ticklish thing." But it produced some important results, such as saving the Kerguelen cabbage and the important grasses that were sent from the Falkland Islands.

A National Collection: The Wilkes Expedition

The United States Exploring Expedition, traveling between 1838 and 1842, often referred to as the Wilkes expedition or the U.S. Ex. Ex., was much more concerned with surveys of the Pacific and exploring American interests on the far side of their huge continent than in an Antarctic descent. But the Antarctic adventure appealed to the ambitious plans of the expedition, and they traveled south, sighted land once, and then returned north. From Antarctica they went to Sydney, New Zealand, Fiji, Samoa, and Hawaii. They then mapped what would become Oregon and Washington, sent out an exploring party to explore the Columbia River inland, and also mapped San Francisco Bay. They traveled back to the American east coast by way of the Philippines, Singapore, and Cape Town. They were the first exploring expedition ever sent out by the United States and the last naval ships under sail to circumnavigate the globe. The American expedition was much larger than either the French or the British: the Americans had six ships and 346 men, including nine scientists, and took with them the largest collection of scientific instruments ever sent with an exploring expedition.[24]

One of those instruments was the Wardian case. Most of the plants, Wilkes wrote, "were transported in Loddiges cases which afforded full protection to them whilst passing through the various zones we necessarily had to go through."[25] It is worth noting that he refers to Loddiges, not Ward. The Wilkes expedition arrived back home with many live plants.

The success of the plant collection rested almost entirely on the wide shoulders of a Scot, William Dunlop Brackenridge. He was six feet tall, strong and resourceful, often direct in his comments but generally kind at heart. He collected many specimens and made up for the inexperienced lead botanist, William Rich. Brackenridge was more experienced in handling exotics than anyone else on the expedition. Before immigrating to America, he worked in Edinburgh as the head gardener at the well-known Canonmills Cottage. Here he cultivated many exotics, including araucarias, acacias, and wisterias. In 1834 Brackenridge left Scotland and worked under Christoph Friedrich Otto at the Berlin botanical gardens. Three years later he made his way to Philadelphia to work for the nurseryman Robert Buist, also a Scot, and famous for cultivating the poinsettia. Interestingly, Buist was close friends with James McNab, who had visited Wellclose Square many times to master the design of the Wardian case. At Buist's suggestion, Brackenridge was taken into the Wilkes expedition to look after living collections. Owing to his robust character and experience, he assumed a very large role in the expedition's botanical results.[26]

The most captivating plant that Brackenridge arrived back with was the California pitcher plant (*Darlingtonia californica*). On the night of 2 October 1841, after a long ascent up Mount Eddy, Brackenridge and the explorers on the inland expedition were visited by a group of Native Americans, who "conducted themselves with great propriety." The next day Brackenridge collected the California pitcher plant in marshes at the headwaters of the Sacramento River. This carnivorous plant from the west captured the curiosity of the public when it was brought back to Washington.[27]

The expedition returned with a large collection of dried specimens, seeds, and live plants. The botanical exploits were one of the most important results of the expedition. While there were already collections in the United States, the contribution of the Wilkes expedition was significant. The herbarium offers a good example. With the landing of the expedition the national herbarium collection doubled overnight. It numbered almost ten thousand individual species, and there were five sets of each species. Many of the duplicates were distributed to institutions across the country and swapped with other naturalists around the globe. There were also 684 species of plants carried home as seed as well as 254 species of live plants, with the number of actual plants approximately double that.[28] The live plants were carried in Wardian cases and came from the final phases of the journey.

The United States was not as well equipped as Britain or France to receive the large collections brought back by the expedition. Originally it was thought that the collections would be stored in the Philadelphia Museum, but by the time the expedition returned home plans had changed—surely such a symbol of state science needed to be shown in the capital. The collections were lodged at the newly built Patent Office in Washington, DC. "Patent Office" may be a misleading name for an institution with a museum and scientific outreach. In the United States, when a person applied for a patent on an invention, they were also required to submit a model of the invention to be placed on display in the museum. Also, at the time there were few scientific institutions in the capital; the Patent Office functioned as a quasi-scientific society that had regular publications and a library in which to host researchers. The collections of the expedition became a main attraction in Washington. The collections were on display at the Patent Office from 1842 to 1857; the ethnological and natural history collections became the founding collections of the Smithsonian Institution. Among its many branches is the United States National Herbarium, where the thousands of pressed plants from the expedition are now deposited.[29]

Brackenridge continued to care for the live plant collection after returning to the United States. When the plants arrived in Washington there was no greenhouse for them, so he personally saw to them until one was constructed on the lawns behind the Patent Office. He was appointed horticulturalist in charge of the living collection and served it for another decade. Because many plants had already been lost, Brackenridge was given strict orders on how to disseminate the ones that remained. At times he even had to deny senators, the first lady, and other amateurs from helping themselves to cuttings.[30] However, Brackenridge was permitted to exchange duplicates with other institutions and nurseries around the world.[31] With this trade the collection grew rapidly. In a short time Brackenridge could report there were "about 1,100 plants in pots."[32] In 1845 the editor of the *Magazine of Horticulture and Botany* visited the greenhouses and reported that "a great accession has been made to the collection, through the untiring exertions of Mr. Brackenridge." By this time the greenhouse had grown to triple its original size. The *Daily Union* reported on 26 June 1845: "His greenhouse is, in our estimation, one of the lions of Washington." Brackenridge's careful tending and judicious exchange of plants enabled the collection to grow into a mark of national pride.[33]

In 1850 Congress approved $5,000 for the relocation of the greenhouse and all the live plants to form the new botanical garden on the west side of the Capitol grounds. These were the founding collections of the United States Botanic Garden in Washington, DC. It is here, at this foundational moment, that we see the great impact of the Wardian case in the United States: it was the prime means of transporting many of these plants from foreign countries.[34]

The effect of the Wardian case is clearly visible in the Botanic Garden in Washington. Today there are at least four plants still growing that were brought back from or are descendants of plants from the Wilkes expedition: the jujube tree (*Zizyphus jujube*) growing in Bartholdi Park, although the original was lost in a thunderstorm in summer 2011; the vessel fern (*Angiopteris evecta*) growing in the jungle, a clone of the one tended by Brackenridge; the Queens sagos (*Cycas circinalis*), with both a female and a male palm in the gardens, both of which came here with the expedition; and the ferocious-looking blue cycad (*Encephalartos horridus*), which Brackenridge collected at the botanical garden in Cape Town on the final leg of the journey.

In November 1842 Brackenridge gave his report on the "botanical department" for the Wilkes expedition. It was a recap of what they had achieved and a statement on the possibilities of the small greenhouse whose virtues he was putting forward for the benefit of the nation. He saw the great potential of plant exchanges: "A nucleus once formed, with a gradual accumulation of stock and a steady perseverance in its support and furtherance, we might, at some not very distant day, vie with the most celebrated establishments of the same kind in Europe." It was a prescient statement that would take another fifty years to be realized, but as we will see in chapter 10, once the Office of Plant Introductions was set up in Washington, it would set the tone of global plant transfers into the twentieth century.[35]

The United States Exploring Expedition is an often-forgotten episode in the history of American exploration. The live plants that were moved in Wardian cases are one example of how the Wilkes expedition served as an important moment in the development of many of the United States' scientific institutions.

Over the Oceans

The botanical legacy of the three expeditions continued for many years. The Wardian case was an integral technology for expeditions and sending plants across oceans. In the decades that followed, many scientists on government-sponsored expeditions were successful in using the Wardian case. Following his first journey, Joseph Hooker went to India, and he employed the cases to send to Kew many species of plants, in particular rhododendrons, that kick-started a thirst for the showy flowers.[36] In the United States a flow of scientific connections was set in motion. Although the case was used on further expeditions, it was the collections of the Wilkes expedition that fed an emerging scientific enterprise. Asa Gray led the project to synthesize the Wilkes expedition's vast botanical collection by seeing many of its publications into print. Brackenridge sent seven boxes of plants from Washington to Cambridge. These not only allowed Gray to complete his publications but were also planted out in the new botanical garden at Harvard.[37]

The use of Wardian cases by the d'Urville, Ross, and Wilkes expeditions led other ships to carry them. The latter two were the last expeditions to travel fully under sail. In 1838, just after all these expeditions set sail, the steamship *Great Western* made the transatlantic crossing in a record fifteen days. It then commenced regular services. The development of steam travel allowed not just people but also plants to move faster, farther, and more frequently. At the same time there was a high demand for new and useful plants by nurseries and botanical gardens. The Wardian case filled an important technological niche in a rapidly expanding global transport system.[38]

5

....

On the Move

It is easy to forget that plant products sustain our lives, from the food we eat to the clothes we wear. The Wardian case moved many useful and valuable plants and helped to establish a number of commercial crops in French and English colonies. The dwarf Cavendish banana was moved from China via England to the Samoan islands and spread throughout the region as a significant crop.[1] When the Frenchman Henri Lecomte was charged with setting up gutta percha plantations in the French colonies of the New World, he took with him plants safely packed in Wardian cases. The importance of the mango for Queensland, Australia, also relied on the Wardian case, with flavorsome varieties arriving from Bombay, Calcutta, and Java. Often when plants were moved beyond their home range, they made a significant economic impact on local agricultural production. This chapter will focus on two of the most important crops in the nineteenth century—tea and cinchona.

The movement of plants was a highly complex procedure, as can be seen clearly in the cases of tea and cinchona, which began to be transported in the mid-nineteenth century. The Wardian case was used by Robert Fortune to ship tea plants to India and became a pioneering example of the case's usefulness in transplanting valuable crops. The transplantation of cinchona from South America to Asia is far more complex and shows both the length of time it takes for an agricultural crop to achieve prominence and the manipulations of nature that were needed to achieve success. By successfully moving important agricultural crops, not only did the Wardian case further establish its utility as a means of moving plants, but its effects were long-term, since many of these plants are still of agricultural importance.

Robert Fortune and Tea

In the eighteenth century Europe began to develop a thirst for Chinese products. This went hand in hand with China's becoming a flourishing market for British products. Traders, surgeons, and ship captains stationed in Canton and Macao fed this growing market with products from the east. Of all the products from China, tea was the most desired. Following the First Opium War (1840–42), five ports were opened to British trade, and Hong Kong became a British possession.[2] The opening of trade with China following the war allowed foreigners greater freedom to travel there.

The Royal Horticultural Society was one of the first to notice the opportunities and quickly arranged to send a plant collector to China. They chose Robert Fortune, a knowledgeable young gardener who, like many other successful gardeners of the era, was Scottish. Fortune was an accomplished botanist and a robust traveler, and he had a talent for retelling the stories of his journeys.[3] From his very first journey to China he was an avid promoter of the Wardian case. His descriptions of the case were an integral part of showing the distances he traveled and the challenges facing plant hunters. Following his final journey in the 1860s, after nearly two decades of plant hunting, he could still tell his readers that travel in remote places was not the hard part; rather, "the difficulty—the great difficulty—was to transport living plants."[4]

In 1843, as Fortune prepared to leave for China on his first journey, he became intimately acquainted with the Wardian case thanks to George Loddiges and John Lindley, who were both on the committee overseeing the voyage.[5] He left England with three Wardian cases filled with fruit trees and ornamental plants that were intended as gifts for people "who may be useful to you."[6]

On his first trip, Fortune collected ornamental plants in and around China's recently opened ports: Guangzhou (Canton), Xiamen (Amoy), Fuzhou (Foochow), Ningbo (Ningpo), and Shanghai. After nearly three years of travel, Fortune boarded the *John Cooper* in Hong Kong ready to return to London with eighteen Wardian cases. The cases were specially made on-site in China with a sliding glass side, and Fortune personally cared for the plants on the return journey. He often opened the boxes to allow the plants to get fresh air. The plants that traveled with Fortune were successful, most of them arriving healthy at the Horticultural Society's Chiswick garden. Inside the cases were many beautiful plants that have now entered cultivation, including varieties of daphne, magnolia, rhododendron, rose, jasmine, and wisteria.[7]

Following his first journey, Fortune gave a full report on the Wardian case that was later published in the Horticultural Society's journal. So important was the transport of plants that Fortune included the report as an appendix to the second edition of his popular book, *Three Years' Wanderings in the Northern Provinces of China* (1847). To Fortune, tending to a case of plants on a long journey not only afforded "amusement" but also enabled one to "enrich one country with the productions of another."[8]

On his first journey in China, Fortune observed Chinese tea manufacture.[9] At the time most foreigners believed that green tea and black tea (then known as *Thea viridis* and *Thea bohea* respectively) were produced from different plants. Through observations in the north and the south, Fortune showed that they were actually the same plant, what we know today as *Camellia sinensis*. The difference was in the manufacturing process: green tea was dried quickly and thoroughly in a large heated pan until it was green and crisp, but black tea was left with a small amount of moisture so that it could oxidize until its characteristic color was produced.[10] There had been much confusion in the botanical literature over green and black tea, and no westerner had witnessed the procedure for producing either in detail. By collecting herbarium specimens of tea plants from various locations, Fortune showed them to be the same species.

Following his return to London, Fortune was employed as the curator at the Chelsea Physic Garden, the botanical garden of the Worshipful Society of Apothecaries and one of the oldest gardens in Britain. At Chelsea he became friends with Ward, who was a member of the garden committee. Their relationship continued for the next two decades.[11] While Fortune was working at Chelsea, the prospects for a tea industry in India were gathering momentum. After only eighteen months at the Chelsea garden, Fortune was contracted by the British East India Company to go to China, collect tea, and take it to India (fig. 5.1).

Many commentators at the time believed that if tea could be cultivated on land under colonial possession, then the Chinese tea monopoly could be avoided. Native varieties of tea had already been found in India's Upper Assam region, so it was chosen as a suitable location for an Indian tea industry.[12] Because tea seeds are very difficult to transport—they are oily and quickly become rancid and inviable—the best way to move them was as live plants. Given Fortune's experience with the Wardian case, it seemed to him only logical to use the case to move tea plants to the new plantations.

Fortune arrived in Hong Kong in August 1848 and immediately continued on to Shanghai. Much has been made of Fortune's journey inland, particularly his disguise—he traveled in full Chinese dress, including having his head shaved. From Shanghai he went up the Huangpu River (Whangpoo) through Hangzhou and Tanxi and as far as the Sunglo Mountains, which had reputable supplies of tea. On his return to Shanghai he traveled to Jintang (Silver Island), known for its tea growing and manufacture. These collections were dispatched from Hong Kong. The next journey was to the famed black tea–producing district in the Wuyi Mountains (Bohea Hills). In the course of his exploits Fortune dispatched many Wardian cases to India.

After some initial failures sending plants, Fortune decided to experiment with a different way of packing the Wardian cases.[13] In one where he had already packed mulberry trees from Hangzhou, he scattered tea seeds over the soil. He then covered them in an inch-thick layer of soil, watered the mixture, and closed up the case. Upon reaching Calcutta the mulberries were doing well, and the surface was covered with sprouting tea plants. The gardener who received the plants in Calcutta wrote to Fortune, "The young tea-plants were sprouting around the mulberries

FIGURE 5.1 Three men carrying loads of tea for the Sungpan market, Min Valley, China, 1908. Photo by E. H. Wilson. © President and Fellows of Harvard College, Arnold Arboretum Archives.

as thick as they could come up."[14] Fortune used this method for all his future shipments of tea. It was not, however, his own idea; it had been used by many before him and was most likely told to him by Ward.[15]

Fortune personally accompanied the final fourteen Wardian cases to India in early 1851.[16] After filling these cases with tea seeds, he still had seeds left over. Not wanting to waste them, he tried to find space in two other cases that he was preparing with camellias. He created a thick mixture of one part soil and two parts seeds. He then placed this at the bottom of the box and laid the camellias in and around the mixture.[17] To his surprise, many of these seeds germinated. In total on this journey, Fortune transplanted nearly twenty thousand tea seedlings to India in more than sixteen Wardian cases.

As important as transporting plants was the knowledge of how to process the tea leaves. In Shanghai Fortune arranged with a Chinese agent for tea workers to be contracted to work in the Himalayan tea districts. They were led by the tea manufacturer Hoo Yuen Fuh and had a three-year contract that, once enacted, could not be withdrawn from. After three years they could choose to continue or return to China. They were paid $15 per month and received $60 in advance as a ticket to India. There were six tea manufacturers and two canister makers.[18] As the sixteen Wardian cases traveled to India, so too did the Chinese workers. Accompanying these Chinese artisans were numerous "implements for the manufacture of tea": six large and six small drying baskets, four round baskets, six mats, ten sieves, nine large sieves, three pans like those used in Stung Chou district, stone slabs, and various other kinds of equipment.[19] Transporting plants was useless without the ability to turn them into a desired product, and the Chinese workers provided the expertise required. Both plants and people arrived in India and were taken to the blossoming tea plantations in Assam.

The East India Company contracted Fortune for another period of collecting in 1853–56. This time Fortune used his many Chinese contacts to acquire seeds from various locations. Again, Fortune used the soil-and-seed mixture to send plants in Wardian cases. In all thirty Wardian cases were constructed for this collecting trip, and tens of thousands of plants were shipped to plantations in northern India.[20] In the following years Fortune sent even more plants: in 1854 and 1855 the tea plants sent by Fortune to the tea districts occupied 131 Wardian cases,[21] a huge investment not only in labor to construct the cases, but also in terms of the amount of glass used to construct them and of soil required. The tea transplant from China to India amounted to one of the largest shipments of Wardian cases ever at the time.

The tea that Fortune sent never did as well as the local varieties of tea. At the same time, it was also discovered by botanists in India that the Indian species of tea was actually the same plant as the Chinese, although they were sufficiently distinct to be divided into two varieties (*Camellia sinensis* var. *assamica*). *C. assamica* was well suited to cultivation in India. But Fortune's tea project had important long-term effects. The Chinese tea manufacturers provided a knowledge base for the new industry, as did their instruments and methods. The first experiments with tea plantations were carried out in Assam well before Fortune had brought tea from China. But owing to Fortune's enthusiasm the tea industry took on new interest particularly for foreign investors and created a vision of wide-scale tea plantations in India. Today the plantations in the Kanga Valley are renowned for their green teas, which are claimed to reach back to stock that was collected by Fortune in China; and the special black teas from Darjeeling are also produced from Chinese-origin stock that can be traced to Fortune's imports in Wardian cases.[22]

By the 1850s tea manufacturing in India had spread to areas beyond Assam. There were the sub-Himalayan districts of Garhwal and Kumaon, Chittagong (now in Bangladesh), Darjeeling, Cachar, and Sylhet.[23] In one region that received many of Fortune's plants, Garhwal Himalaya, in and around the Saharanpur Botanical Gardens, between 1844 to 1880 tea cultivation expanded from seventy acres to more than ten thousand.[24] By the 1870s it had spread to Ceylon (Sri Lanka). Today both India and Sri Lanka are major global tea producers.[25]

Over the fifteen years that he worked in China, Fortune garnered a reputation as an expert in Chinese tea. In 1857 the commissioner of the US Patent Office, Charles Mason, contracted Robert Fortune to collect tea. Mason's plan was to send tea seeds in tin boxes. Fortune told Mason

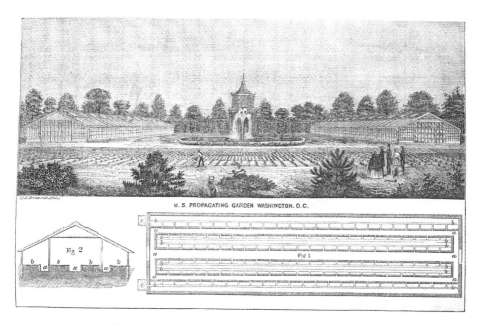

FIGURE 5.2 US Propagating Garden, Washington, DC, 1859. The glass houses pictured on the left and right were built to receive the tea plants sent in Wardian cases to the United States by Robert Fortune. From *Report of the Commissioner of Patents: Agriculture* (1859).

that the best method was to send live plants in Wardian cases. As the seed merchants Charlwood & Cummins, who acted as intermediaries in the project, told Mason: "Mr. Fortune assures us this way only (that is by the seeds being placed in soil in Ward's cases) is there any chance of success."[26]

In 1858 Fortune traveled to China on his fourth expedition. Thanks to the many contacts he had developed in China, what once took him many years to collect now only took him a season.[27] He packed the seeds in Shanghai and sent them to Washington. In total Fortune sent more than thirty thousand plants and nearly twenty Wardian cases (fig. 5.2). In April 1859 they arrived in Washington in "excellent order."[28] To receive the plants, the Americans constructed two large greenhouses on five acres of public land on Missouri Avenue between Fourth and a Half and Sixth Streets. In the United States there were many requests for tea seedlings, particularly from farmers in the South. Many were distributed. But soon afterward, with the outbreak of the Civil War, the tea project was abandoned.

For Fortune, the Wardian case was an important tool for moving plants. Often plants had to travel a great distance, but sometimes the plants needed were right under the botanist's nose, such as the native Indian variety of tea. Most striking about Fortune's use of the Wardian case is that the possibility of moving an actual plant led to an important transfer of knowledge between China and India. Moving both the product *and* the people who manufactured it paved the way for an Indian tea industry.[29]

Grafting Cinchona

Malaria has been a scourge for centuries if not millennia. The disease was most devastating in tropical and subtropical regions—not only Africa, but also China, India, South America, and Southeast Asia. The coastal deltas and swamps in Europe also had major outbreaks. Among the many remedies that were tried in the seventeenth century, the one that garnered the most attention was *quina*.[30] Today we know this drug as quinine. The alkaloid quinine, discovered and first used by local people, was sourced from the ground-up bark of the cinchona tree, which grew in the high Andean forests. Quinine prevented fever, and its healing properties became widely known.

In the mid-nineteenth century cinchona was probably the most sought-after economic plant on the planet. It was of much higher value than tea, bananas, and most others. The British and Dutch were very keen to appropriate cinchona from South America, and the Wardian case was a key tool in their efforts. The work of Fortune in moving tea from China had shown imperialists that moving large quantities of valuable plants was possible. A few years later Clements Markham, who led the British efforts to transplant cinchona, sat down with Fortune to learn the best ways to use Wardian cases to move the plant from South America to India.[31]

As empires spread, cinchona's importance grew dramatically. It was the chosen preventative used by the British, Dutch, and French.[32] It was used by explorers traveling in tropical locations; by soldiers, traders, and government officials posted in malaria-affected areas; and later to supply local populations with a preventative. A dose of powdered quinine was a daily affair for colonists in tropical regions. But so bitter was it that people often consumed it mixed with alcoholic spirits or wine.

The British and Dutch governments decided to introduce cinchona from its native South American locations to plantations in their colonies. It is one of the most infamous acts of biological espionage in world history.[33] A whole cast of gardeners and plant hunters were enlisted to find plants and deliver them to the Asian colonies. The cinchona was moved from South America as both seeds and live plants, the latter primarily by means of the Wardian case. The case was used by the two leading figures employed in the transplant: Markham for the British and Justus Karl Hasskarl for the Dutch. In the 1860s, soon after the imperial powers transplanted cinchona, South American nations banned the removal of the plant from its native locations. But it was too late; the seeds had already been planted in Java and India.

Hasskarl was a German-born botanist working for the Dutch. In 1852 he was contracted by the Dutch government to journey to South America to collect cinchona seeds and plants for the purpose of starting plantations in the Dutch Indies. Arriving in 1853, he went from Lima to the Andes, where he collected many varieties of cinchona plants. Traveling under an assumed name, he made numerous collections on the Bolivian frontier and packed his cinchona plants into twenty-one Wardian cases. In late 1854 he set sail for Java with the cases and a large quantity of seeds.[34]

The cinchona plants survived. There were a number of known species, including a new species of cinchona, *Cinchona pahudiana*, named for the governor-general of the Dutch East Indies, Charles Ferdinand Pahud. By 1860 there were nearly one million plants of *C. pahudiana* in plantations on Java. However, different species of cinchona have varying amounts of quinine in their bark, and when the bark of *C. pahudiana* was tested, it was found to contain virtually no quinine. The whole project was such an embarrassment that the failure was aired in the Dutch Parliament in the early 1860s. By 1865 the Dutch were close to abandoning their cinchona plantations. The disaster was blamed on a number of factors, but all of them came back to the lack of holistic knowledge of either the plant or where it grew.[35]

Five years after the Dutch debacle Markham led the British effort to acquire cinchona. The project was spread over three locations in an attempt to collect different varieties of the plant.[36] Markham traveled to Bolivia with his wife, Minna and collected plants with the help (and significant labor) of Robert Weir, who was recommended to Markham by the Veitch nursery. Richard Spruce and Robert Cross were sent to collect in Ecuador, and G. J. Pritchett was given broad license to collect whatever varieties he could find in Peru. While plants were being collected in South America, a site for a plantation was selected and cleared near Ootacamund (today Udhagamandalam), Madras, in southern India. At the same time a new propagating house at Kew Gardens was erected to serve as the halfway house for the many plants on their way from South America to India. The logistics of the project were extensive.

Thirty Wardian cases were constructed at Kew, packed as flat packs, and sent around Cape Horn to ports in South America ready for collectors (fig. 5.3). They were three feet two inches long, one foot ten inches wide, and three feet two inches high. Filled with soil and plants, each weighed 336 pounds (152 kilograms). Fifteen went to Ecuador for Spruce's collection, and fifteen went to southern Peru for Markham's. Pritchett was originally sent out to gather seeds but had six cases made by a carpenter in Lima to take his plants to India.

Together with the gardener, Weir, Markham traveled from Islay over the Andes to the cinchona regions around La Paz, Bolivia, and collected plants and seeds. They arrived back in Islay (today Mollendo, Peru) with plants enough to fill the fifteen cases. The cases were packed with 458 C. calisaya plants. Before setting off, the plants appeared in excellent health. The cases left on the steamer on their way through Panama and a very hot journey through the Red Sea.

Markham took a different route from the plants but met them in India. He arrived at the port city of Calicut (today, Kozhikode), and soon afterward a canoe trip upriver set out on the thirty-three-mile ascent into the mountains, where the newly established plantation in Ootacamund

FIGURE 5.3 The Wardian case, one of the world's first flat packs. From "Hints for Collectors," *Bulletin of Miscellaneous Information (Royal Botanic Gardens, Kew)* 3 (1914). © The Board of Trustees of the Royal Botanic Gardens, Kew.

was being carved out. The fifteen cases were carried through the rain and cold by 150 Indian laborers. It was difficult work, Minna Markham recorded in her diary: "The poor coolies who had serviced the cases from the low countries, felt the cold terribly and cowered about shivering, holding lighted sticks to warm themselves."[37] The plants arrived in Ootacamund in poor condition. Initially the blame fell upon Weir, who had packed them in Peru. Minna Markham wrote in her diary: "Mr. McIvor says that not even a willow could have lived in such mud!"[38] Later Markham blamed the high temperatures they encountered when the plants crossed the Red Sea as the cause of the failure. Although Markham failed, the success of other collectors allowed him to claim the entire project a triumph.

Spruce and Cross's collections were taken from the cinchona forests around Limón, Ecuador. There Spruce and Cross propagated cuttings of red-bark cinchona, *C. succirubra*. When the trees flowered in August 1860, one of the local laborers climbed a high cinchona tree to get seeds. He broke off the loose branching clusters of flowers and dropped them onto sheets laid out on the ground. Within a few months, and with much help from local laborers, they had a modest supply. They then moved the plants down to the river, where they could travel to the port. At the river, the flat-packed Wardian cases were waiting, and in late November 1860 Spruce worked with a local carpenter to assemble them: "I found a negro carpenter to aid me in putting the cases together, but he was used to only rough work, and to nails of the largest calibre, so that if I had not put in most of the nails with my own hands, the cases would have been split in pieces."[39] Many colonialists seemed not to understand that without the help of local people, they would never have had any plants to pack in the first place.

Once constructed, the cases were filled with 637 plants that had come from the hills. Inside the cases Cross even managed to squeeze in an interesting white-flecked orchid collected from the banks of the river. The seeds and plants collected by Cross were well packed and had cool weather for their journey, and he accompanied them to India. They stopped briefly at Kew, where six plants and the orchid were taken out and then sent on to India for the plantation. The surviving 463 cinchona

FIGURE 5.4   Seedlings of *Cinchona succirubra* on arrival at the Botanical Gardens, Ootacamund, in southern India, 9 April 1861. This is the earliest known photograph of a Wardian case. © The Board of Trustees of the Royal Botanic Gardens, Kew.

plants were photographed on their arrival in Ootacamund (fig. 5.4); it is the earliest known photograph of Wardian cases.

C. succirubra became the favorite variety of William Graham McIvor, who oversaw the cinchona plantations at Ootacamund. It was one of the hardiest of all varieties and appeared to thrive in the Indian environment. This attribute became even more important in the coming decades. From the initial plantation, others were established across India. As more plants were propagated, the Wardian case was used more and more. Hundreds of cases were sent to people wanting to set up cinchona plantations.[40] On one transport in 1862, McIvor sent thirteen cases to Sikkim. The journey was a difficult one; on one section it took two days to travel twenty-four miles. The delay was caused by the "inefficient coolies," as the colonialist overseeing the transplant told his superiors at the India Office. He went on to explain why the workers were so slow: "Many of them were children of 12 and 14 years of age, who were not even tall enough to lift the cases from the ground."[41] To be clear: McIvor was using children to transplant cases weighing nearly 330 pounds (150 kilograms). Two of the cases did not get very far: they were left on the side of the road in the burning midday sun, with the young people "lying by them completely exhausted." At all times the Wardian case witnessed the uneven distribution of labor and the horror of emerging agricultural economies.

Within a few decades millions of cinchona trees had spread over India and Ceylon. The variety that Spruce and Cross brought back in Wardian cases, *C. succirubra,* was the most common one planted in the subcontinent. It was largely a failure. While it contained other important alkaloids valuable to pharmaceutical companies, its quinine content was too low to be of commercial value, as the market shifted toward purer sources of quinine from different trees. The British persisted for decades, but the cinchona plantations became unprofitable. By the 1880s millions of cinchona trees were felled in India and the bark dumped on the market. The market for cinchona collapsed, and Indian planters turned to growing tea. The British experience was similar to that of the early Dutch venture. "But in India," reflected the cinchona expert Norman Taylor in 1945, "with characteristic British tenacity, the government obstinately clung to species and methods long proved disadvantageous."[42]

Interestingly, both the British and the Dutch were in regular contact regarding the science of cinchona.[43] In 1861 Thomas Anderson, then director of the Calcutta Botanic Garden, was sent to Java to exchange cinchona plants. When he returned, he had seven cases filled with plants from Java.[44] The cases were transported 170 miles from the Tijaneroian plantation to Batavia by Javanese local workers, who carried them at night and rested with them during the day under fig trees in the villages along their way. The following year, with the specimens of *C. succirubra* progressing well in India, the British sent the Dutch plants of this new species to add to their stocks. The transfer of seeds, cases, and knowledge between the British and the Dutch continued for the next decade. In general, both knowledge and specimens of cinchona were openly transferred between the scientists.

While the British persisted with *C. succirubra*, the Dutch commenced with another variety. The story of the plant that came to dominate the market has far less expeditionary grandeur than either the Dutch or British efforts, but it still carries its own intrigue.

As a British trader and longtime resident in the Peruvian highlands, Charles Ledger had traded in cinchona bark and even knew where much of it grew. His first claim to notoriety was a failed attempt to export alpacas to Australia. On many of his alpaca journeys Ledger was helped by his local guide and friend of twenty-five years, Manuel Incra Mamani. On one of their alpaca expeditions Mamani told Ledger of an area where the cinchona with the best bark grew and promised to secure seeds.[45]

One day in 1865 Mamani turned up on Ledger's doorstep with more than thirty-five pounds of cinchona seeds collected from areas with the highest-yielding bark. Ledger packaged the seeds and sent them to his brother George in London to sell to colonial traders. George shopped them around to all the leading botanical and imperial circles but got relatively little interest from the British. At the time most of the British involved in the cinchona project thought that Markham and his team had collected enough cinchona to kick-start their plantations, so they neglected Ledger's seeds. They also believed that the *C. succirubra* was better suited to the climate in India.[46] However, he eventually sold his seeds to the Dutch Counsel General in London.

While the Dutch waited in Java for the seeds in their small nursery to grow, Ledger asked Mamani to go on another expedition for seeds.[47] After another good collecting season, Mamani was on his way back when he was stopped by authorities. The seeds were found, and he was imprisoned. (Exporting cinchona seed had been outlawed in Bolivia soon after the British took seeds and plants.) At the jail he was severely beaten. After three weeks he was released, only to die a few days later from his injuries.

In Java a small nursery was set up, and about twelve thousand of Mamani's seeds were tended. By this time Karl van Gorkom—a practi-

FIGURE 5.5 Grafting *C. ledgeriana* scions on *C. succirubra* rootstock at Tijaneroian plantation, West Java. Collection Nationaal Museum van Wereldculturen, Image TM-60018886.

cal agricultural scientist well aware of the needs of planters and quinine producers—had taken over Java's government plantations. As the plant grew it was named a new species, *Cinchona ledgeriana*. A few years later a young chemist from Holland came to test the quality of the bark. To his surprise, the quinine content of the bark was as high as 13 percent. It was a major commercial breakthrough. Before this, barks with 3 percent quinine were considered exploitable, and those with 5 percent were very good.[48] With such high yields of quinine, these new trees would garner a huge swath of the quinine market. But there were many challenges to getting the plants to grow in plantations.[49]

Gorkom soon realized that *C. ledgeriana* thrived in the rich soil of newly cleared virgin forests. When the plants were deposited in less fertile soil, they suffered and grew stunted. Under the guidance of a horticulturalist, it was found that grafting scions of *C. ledgeriana* onto rootstock of the hardy *C. succirubra* would solve the problem (fig. 5.5).[50] *C. ledgeriana* continued to thrive in these conditions, and the hybrid seedlings were distributed to plantations all over Java.

By the 1930s the Dutch were exporting two million pounds of bark annually and controlled nearly 90 percent of the market. It was worth millions to their colonial agricultural economy. The British largely turned their attention to tea plantations, and the Dutch, through much persistence, had their own cinchona plantations. It was not until World War II, when Japan took Java, that the monopoly in quinine production was unsettled. "Chimera" is the label botanists give to graft-hybrid plants such as the plantation stocks of Dutch cinchona. This chimera is formed from high-producing quinine stock collected in the highlands in Bolivia by Mamani and from *C. succirubra* stock that had traveled in Wardian cases on many journeys.

We see in this complex story that the Wardian case was an important technology for transporting plants. Many of the transplants were failures. Markham's plants did not survive the journey from South America, and Hasskarl's and Spruce's plants had low yields. Although it was the seeds of Mamani and Ledger that ultimately launched the Dutch cinchona plantations, the inadvertent transplant of the hardy variety of *C. succirubra* was an important ingredient in the Dutch success. The knowledge about and varieties of cinchona species that were moved is a forgotten part of the story. That nature was so manipulated and standardized through agricultural science was only possible by moving a vast variety of plants.

6

House of Ward

Nathaniel Ward's door was always open. Whether welcoming plant hunters such as Robert Fortune or leading botanists such as Joseph Hooker, Ward's home was a sanctuary for naturalists. One could always be assured of a meal and good conversation. There were also many new and interesting plants to inspect, either under a microscope or growing under glass. Ward's friends and contacts continued for years to send him plants. Joseph Hooker described Ward's home: "During the whole period that I knew him, and I believe for many years before, his hospitable house first in Wellclose Square, and latterly at Clapham Rise, was the most frequented metropolitan resort of naturalists from all quarters of the globe of any since Sir Joseph Banks' day."[1] The success of Ward's invention for moving plants cannot be separated from his networks. The very name of the box, the Wardian case, cannot be fully grasped without visiting his hospitable house.

In 1849 Ward left his small, dark house at Wellclose Square in the center of London for the fresh air of Clapham Rise. Located in southwest London, Clapham was three and a half miles from the city and about a thirty-minute horse-drawn-cab ride from London Bridge. His son Stephen married and took over the doctor's practice and the Wellclose home. Ward kept some of his patients and even took on new ones in the

MR. WARD'S FERN-HOUSE.

FIGURE 6.1 Interior view of Ward's greenhouse at Clapham Rise, 1851. From *Gardener's Magazine* (1851).

vicinity of Clapham. After the move he had more time for botanizing. At Clapham he turned his modest-sized plot into something wonderful that could lavishly express his worldwide collections. Ward set to work on the garden immediately. One of the first features he worked on, following the fashion of the times, was a rockery. A North American ground pine (*L. dendroideum*) sent to him from Boston by Asa Gray was one of the first plants to go into the ground.[2] He also set up heated greenhouses and specially constructed cases for growing varieties of Australian, Indian, North American, and even Arctic plants. Ward named his Clapham house The Ferns, for one of his favorite plants (fig. 6.1).

The garden grew wildly. After a decade at Clapham, Ward could boast that "my own garden is most luxuriant—[William Henry] Harvey says it ought to be called the Wilderness—a name it certainly deserves."[3] There were always more plants and more growth. Only after fifteen years could Ward declare it almost complete. But, like most gardens, it would never be finished. Ward went on working his garden, and his friends continued to come and visit and enjoy his hospitality. One friend described the house as cheerful and interesting, with plants growing everywhere that delighted both the visitor and the cultivator.[4]

Entertainments at Ward's house were frequent, at their height even weekly. On these occasions his many scientific friends came to see him and his plants. Often at these parties "many a country and colonial naturalist was introduced for the first, and too often the last time in his life to some of the most eminent naturalists of Europe."[5] In the mid-nineteenth century a house like Ward's could be visited by amateurs and professionals alike, all connected by their passion for plants.

As the house at Clapham blossomed, so too did the use of the Wardian case. In 1853 Ward wrote to Gray: "The closed cases are progressing—or going ahead in a way which would be most gratifying to me, could I spare a little more time and means."[6] By the mid-nineteenth century Ward's parlor cases were all the fashion in middle- and upper-class Victorian homes. In particular the fern craze, *pteridomania*, gripped Britain and Europe. With Ward's cases botanical treasures could be brought inside. A wealthy London lady had fifty cases constructed to donate to an event. Like many others, she called upon Ward to advise and help her in planting the cases. And just as Ward was called upon for knowledge, at Kew William Hooker was overwhelmed with requests for exotic specimens of the beloved plant. "A shipload of ferns would hardly suffice the present demand," he told Ward.[7]

A Network of Friends

For an amateur naturalist, Ward was widely connected. Some of his closest contacts were the Hookers (father and son) at Kew; Asa Gray in Boston; William Henry Harvey at Trinity College; Robert Wight, a pioneer of Indian botany; and Edward Cooke, the famous illustrator and inventor of the aquarium. One of the impressive aspects of Ward's connections were how he maintained them over decades. He met the young Asa Gray in 1839 and remained in correspondence with the American for nearly three decades. It was the same with Joseph Hooker—their correspondence also spanned three decades. Ward's Clapham house was not far from the boy's school that Joseph Hooker's and Charles Darwin's sons attended. Their young sons could often be found taking advantage of Ward's hospitality and being instructed in botany.[8]

Ward gave his time to all manner of people, not just eminent scientists. He continued to maintain a good relationship with Charles Mallard, the ship captain who delivered the first Wardian case to Australia. When Mallard settled on land in northern Australia, on the Darling Downs grazing district inland from Brisbane, Ward asked Mrs. Mallard to collect for him, and in response she sent him cases of interesting Australian plants. Sifting through the plants, Ward found a number of new ones; he packed these and sent them on to Gray in Boston for identification. In a note accompanying the plants Ward wrote, "It would give me the greatest pleasure could you (if you find anything new) give her the credit of it."[9] Ward created an extensive network that saw letters, people, and plants traveling around the world.

A list of Ward's correspondents would be a who's who of nineteenth-century natural history. Augustin Pyrame de Candolle was a Swiss botanist and plant geographer who laid the foundations of modern biogeography. Ward dined with him in 1816, after which he maintained a correspondence with him and visited him in Geneva. In later years he entertained Candolle's son at Clapham. The young German Theodor Vogel, the collector on the Niger expedition in 1841, during which he died of dysentery. Ward was also friends with William Stanger, who traveled to Africa with Vogel but returned unscathed. Both Vogel and Stanger were given a send-off at a party at Ward's house and were instrumental in taking the Wardian case on the expedition. Stanger would go on collecting and using Wardian cases, such as when he traveled to South Africa and sent Ward specimens. There were the Americans: Daniel Cady Eaton at Yale; the New Yorker Jacob Whitman Bailey, pioneer of the microscope in America; and the Ohio bryologist William Starling Sullivant, who maintained a relationship with Ward and sent him many specimens. Although happy to receive species from the Midwest, Ward was concerned with the number of specimens Sullivant was distributing to scientists; "Is he not making too many species?" Ward once asked Asa Gray.[10] In Australia, Ward was in contact not only with leading scientists such as Ferdinand von Mueller in Melbourne but also doctors working on the goldfields such as Thomas Le Gay Holthouse, who as a young student in London had spent many hours studying at Ward's house. There was also the widely influential and mostly self-taught geologist and fossil hunter Charlotte Murchison, who also attended Ward's house.[11]

Of course, such a network was not uncommon for scientists. The networks of Gray and Joseph Hooker were much larger, but for Ward, an amateur whose primary profession was medicine, this is an impressive list of friends from around the world. His engaging and outgoing personality enabled him to promote his cases to others. In 1853 he wrote in a letter to Gray: "My garden and cases are beginning to get full of interest to the botanist thanks to you and to other friends."[12] The members of Ward's network advocated for the continued use of the cases he invented. Nearly all had at some point been to Ward's house and seen his cases firsthand.

Banana Journeys

Ward's many and varied contacts led to significant outcomes for the movement of valuable species. Ward was friendly with the missionary Reverend John Williams and advised him on how to best move plants. In 1839, as Williams was preparing to leave on yet another pastoral trip to the Pacific, he heard about the potential value of the dwarf Cavendish banana (in Ward's time it was named *Musa cavendishii*, but today it known as a cultivar of *Musa acuminata*). He asked Ward if the plants would travel in a glazed case to the Pacific. Ward encouraged the transplant.[13] Originally domesticated in tropical Asia and New Guinea by native peoples over thousands of years, the dwarf Cavendish banana arrived in Britain from Mauritius in 1829. One plant found its way into

William Cavendish's greenhouse at Chatsworth, and his head gardener, Joseph Paxton, cultivated it. By 1838 Paxton reported that he had a hundred plants to distribute. Paxton gave Williams a plant to pack into his two Wardian cases to take to the South Pacific. "I have just heard that the Musa cavendishii has been introduced into the Navigation Islands by means of the glazed cases and that it promises to be a most valuable addition to their vegetable stores," Ward wrote excitedly to William Hooker. So thrilled was Hooker about the banana transplant that he requested further details from Ward.[14]

Six months after leaving London, the banana was carried ashore on the small South Pacific island of Upolu, today part of Samoa. When Williams unpacked the Wardian case he thought the plant almost dead and did not give it much of a chance of survival. William Mills, the missionary based at Upolu, was able to get one plant to grow and planted it in the mission's small garden. It flourished, and by 1840 there were more than three hundred banana trees and thirty suckers.[15] The story of Williams was not so fortunate. Although it was widely known that local populations of the South Pacific islands were hostile to foreigners, a few months after arriving in Upolu Williams traveled to the New Hebrides (Vanuatu), where he and his assistant were clubbed to death in retaliation for attacks by earlier sandalwood collectors. Williams's death made him a martyr, and his story and that of the banana transplant were often told together. As missionaries liked to tell it, the arrival of the missionaries brought many fruits both civilizing and economic.[16]

The banana is nutritious and flavorsome and easily cultivated by cloning—meaning that the missionaries could simply transplant suckers of the first banana tree and take it to other islands. The banana spread throughout the Pacific, from Tahiti to the Torres Straits and even as far as Hawaii. The dwarf Cavendish banana is the most common type of banana available in supermarkets today.[17]

Ward was fond of telling the story of Williams and the banana transplant. While entertaining Williams's wife, Mary Chawner, she recounted the banana story to her host, and it was from her description that he was able communicate the success of the Wardian case in the banana transplant. Ward was in close contact with other missionaries in the

Pacific. He encouraged them to get the native people to collect botanical specimens and send them back to London. For Ward, the utility and necessity of understanding the natural products of a location were "not only . . . a means of adding most materially to the comforts both of the missionaries and the natives, but . . . the most efficient way of leading the infant savage mind to the knowledge of a great just cause."[18] This was the thinking of the mid-nineteenth century. Botany was not only a great and just cause; it was a way to colonize.

On Glass

Together with a number of scientists and medical practitioners, Ward played an important role in abolishing the glass tax. Starting in the middle of the eighteenth century, glass was taxed based on its weight. This practice continued for nearly a century. In the 1840s a wide section of professionals, from scientists to medical practitioners, began to argue against the tax. Considering that many of its readers had their own greenhouses, it was not surprising that the *Gardeners' Chronicle*, led by John Lindley, was particularly vocal about the matter, and throughout 1844 and 1845 the newspaper ran long articles calling for an end to the tax. The Commission of Inquiry into the State of Large Towns and Populous Districts was one of the governmental bodies that heard opinions on the issue.

Ward was called to give evidence. He gave details from his work as a medical practitioner in the dark, dirty, and heavily populated East End. Ward told the commission that if more light could be admitted to dwellings of the poor, then their hygiene would be greatly improved. But during his testimony questions also turned to the Wardian case. If the tax could be removed, then all people, rich and poor, could have Wardian cases in their home. Pushed on the issue, Ward even told the commission, "There is light enough in the most dirty parts of London to grow plants of the most delicate kind." He even provided an example, the Killarney fern (*Trichomanes speciosum*), which he said had defied cultivation but now, "if placed in a glazed case, will now grow in any blacksmith's shop in London."[19] In arguing for the positive impact repeal of the glass tax would make on the poor, Ward was also able to promote his cases.

Following the recommendations of the Commission, the tax was abolished by the government of Prime Minister Robert Peel. On 22 February 1845, soon after Peel's speech to the House of Commons, the *Gardeners' Chronicle* led with an editorial from Lindley on the issue: "Whatever the opinion of politicians may be respecting Sir Robert Peel's financial measures, there can be no doubt that the Right Hon. Gentleman deserves the hearty thanks of all persons having gardens."[20] Because it was no longer taxed on weight, glass could profitably yet cheaply be made in thicknesses great enough to make the product durable for use in anything from greenhouses to Wardian cases. It was an important technical turning point that led to much wider use of glass in homes and in long-distance plant transportation.

The Great Exhibition

In the autumn of 1850 Ward welcomed one of the duke of Devonshire's gardeners to see his plant cases.[21] It had been over a decade since the young gardener John Gibson had traveled to India and filled Wardian cases with novelties for Paxton's palm house at Chatsworth. This time the gardeners were calling to question Ward about his plant boxes, both the original construction and the newer types of cases he was now using. By this time plans were rapidly gathering momentum for the Great Exhibition of the Works of Industry of All Nations. The Crystal Palace, designed by Paxton and in some ways inspired by his original hothouse in Chatsworth, was under construction in Hyde Park. It would showcase the great industrial and economic exploits of nations all around the globe.

The Great Exhibition was an important moment for the Wardian case. First, it featured the case as one of a number of interesting industrial exploits. Second, with the construction of the Crystal Palace, glass began to be produced and purchased cheaply on a commercial scale. The Crystal Palace was an architectural achievement of the time. Covering almost eighteen acres, the main building was over 1800 feet (563 meters) long and 400 feet (124 meters) wide, and at its greatest height it was 108 feet (33 meters) tall.[22] It was largely made of glass. In total 300,000 glass sheets were used to cover the huge structure. Such a construction would not have been possible without the repeal of the glass tax years earlier. At their height the Chance Bros., who held the glass contract, made 63,000 sheets in just a fortnight during January 1851. The construction of the Crystal Palace made the mass production of glass possible, which influenced the ongoing production of new Wardian cases.

Opening in May 1851 and concluding in October, the Great Exhibition would see more than six million people through the gates. Inside this glass structure were not only the largest collection of manufactured products from around the world but also thriving productions from nature, including a row of large elm trees that already existed in Hyde Park and were covered by the building. The historian David Allen has said that the palace made "the whole country extremely, even excessively, aware of . . . foliage inside glass."[23] The Wardian case appeared at various parts of the exhibition. People could view economic and ornamental plants brought from far off regions. And they could also view curious objects like Ward's glass bottle with a plant in it that had apparently not been watered for eighteen years.[24]

There were tens of thousands of objects displaying the products of many nations. There were false teeth and hydraulic presses, vulcanized tires and unique tapestries. Some attendees seem to have been somewhat surprised that the Wardian case made it into the exhibition at all; one writer for the *Illustrated London News* aptly put it that the visitor "might naturally be at a loss to account for the reason why so uninteresting an object has been sent to the World's Fair."[25] The exhibition included many different varieties of cases. While several of the well-adorned parlor cases were indeed quite curious, in the east gallery there were also plain traveling cases. According to the journalist, Wardian cases, particularly the traveling cases, were as important a part of British industry and worldwide connections as steam engines or vulcanized rubber. With the cases they could send out useful plants that colonists would delight in, and in return other plants could come back. By 1851 not a week would pass without ships arriving in Britain with cases full of plants from "the remotest habitable regions" around the globe. The cases, the journalist went on, "have thus conferred upon us a power of procuring exotic vegetable productions" that could not have otherwise been obtained. The exhibition also displayed many ornamental plants in Wardian cases. Some of the plants shown inside cases were the porcelain flower (*Hoya carnosa*) and the closely associated *Hoya bella*, which was in flower for the exhibition. And in the north transept of the Great Exhibition one could find a case that housed both ferns and cacti.[26]

The Wardian case was an accessible object—everyone could own one. The journalist for the *Illustrated London News* left the final comment for bringing nature indoors: "The cultivation of plants is an occupation delightful in itself, and one that is calculated to afford intense pleasure to those who follow the amusement." Every child in London, proposed the journalist, needed a Wardian case, so that a love of cultivation would grow with them as they grew up. The inclusion of the Wardian case in the Great Exhibition was a major milestone in its acceptance, probably more for ornamental reasons and for its use by amateurs as a parlor centerpiece. But even as the case was admired for its aesthetics, it was not lost on commentators how important this technology was in bringing plants across the oceans.

Second Edition

The widely revered publisher of books on natural history from Paternoster Row, John van Voorst, spent much time at Ward's house. Often it was taken up with staring through the lens of a microscope. Both Ward and van Voorst were founding members of the Microscopical Society, which was formed one evening in 1839 at Wellclose Square. It is not surprising that, following the success of the Wardian case at the Great Exhibition, Ward, with encouragement from van Voorst, began working in his spare time on a second edition of his book *On the Growth of Plants in Closely Glazed Cases.*[27] In August 1852 he finally delivered the second edition to van Voorst. Ward wrote about the challenges of writing in a letter to Gray: "If I had the pen of a ready writer I could make it most interesting, as the subject matter is second to none. I have however done my best and I am vain enough to think that you will consider it an improvement on the last."[28] Ward may have been a very good networker, but he was never a writer. He often retold the anecdote of the well-known explorer of Africa, David Livingstone, who told him that "the labour of writing his book of travels was far greater than the labours of travelling."[29]

The second edition of Ward's book is largely the same as the first. A couple of additions are of interest. There is a new preface, in which Ward spends much time talking about parlor cases but adds with some confidence: "As to the conveyance of plants on shipboard, the plan is now universally adopted, and it is believed that there is not a civilized spot upon the earth's surface which has not, more or less, benefited by their introduction."[30] In chapter 4, on the use of the cases for transporting plants, Ward updated details of plant hunters successfully using the cases on shipboard. The primary example was, of course, Robert Fortune's exploits in China. The letters that made up the appendix were also updated to show further support for the cases, many dealing with their use in hospitals and sanatoriums. One of the most noticeable differences between the first and second edition was the illustrations, which were executed by Ward's daughter-in-law Georgiana Eglinton Ward (née Cooke), wife of Ward's son Stephen. She came from a family of well-known illustrators. Her brother Edward William Cooke, who also contributed illustrations to the second edition, married George Loddiges's daughter Jane.[31]

As he had with the first edition of the book, Ward sent out many copies through his networks to friends all over Britain and Europe, as well as to distant lands where Ward had friends: Australia, North America, and Africa. His friends gave it good reviews. Ward's faithful correspondent in Boston, Asa Gray, agreed with Ward's claim that its use for transporting plants across oceans was "one of the most important practical applications of Mr Ward's discovery."[32] Gray went on to observe that "this mode of conveyance is now universally adopted, and has proved so successful, whenever properly managed."

The early 1850s, with the Crystal Palace and the new edition of his book, were high points in Ward's twilight years. In 1856 his friends arranged for him to sit for the artist J. P. Knight in recognition of the contribution his cases had made to both individuals and the state. As his son Stephen would later put it, it was organized with the "love and friendship of a very large circle of friends."[33] For many years the portrait hung in the Linnean Society (see fig. 1.1).

Scientific *Conversazione*

Ward's lasting achievement in the 1850s, the one that he was most proud of and most invested in, was saving the Chelsea Physic Garden from ruin. Today in most of the English-speaking world we call our drug makers pharmacists, but in the nineteenth century they were known as apothecaries. In Germany they still call their drugstores an *Apotheke*. "Apothecary" has its root in the Greek word *apothéké*, meaning a storehouse of wines, herbs, and spices. In Ward's time apothecaries dispensed anything from aphrodisiacs to antiseptics, and to do so they needed a thorough knowledge of plants. Physicians in the late nineteenth century also needed to understand the natural foundations of their prescriptions. To successfully complete their medical training many physicians had to pass botany exams (a requirement not abolished until 1895). Ward, Joseph Hooker, and Ward's son Stephen all had to pass such exams, and all three would go on to become examiners at the Society of Apothecaries.

Apothecaries needed to be able to identify medicinal plants and compound them into effective remedies.[34] From the thirteenth century on, physic gardens were places where useful plants were gathered together for use by apothecaries. Located in London, the Chelsea Physic Garden was the place of practice for the Worshipful Society of Apothecaries, who were authorized to license medical practitioners. Founded in 1673, the Garden is the oldest botanical garden in London still in existence. For a time in the late eighteenth century, with Phillip Miller at the helm, its large variety of exotic plants made it perhaps the most famous garden in Europe. In 1846, between trips to China, Robert Fortune worked as the curator at the Chelsea garden, and in 1833 Ward became a member of the Garden's board. Quickly spreading his influence, in 1836 he even had the curator make up two glazed cases to send to India.[35] Ward's work

FIGURE 6.2 Scientific *conversazione* attended by nearly seven hundred people at the Society of Apothecaries' Hall, Blackfriars, 1855. From *Illustrated London News*, 28 April 1855.

as a physician and his passion for plants coexisted comfortably at the Chelsea garden, and for three decades he advocated for it to demonstrate scientific progress.[36]

On 4 September 1854 Ward was elected Master of the Society of Apothecaries.[37] Soon after his appointment he started planning two grand parties to showcase science and education as important parts of the Apothecaries' mission. On 7 March 1855 and five weeks later on 11 April, Ward held two of the largest microscopical parties London had ever seen (fig. 6.2). His daughters and the servants at Clapham worked around the clock to send out nearly eight hundred letters of invitation. The evening's entertainment was centered on the microscope—a hundred were set up. He had his friends donate an almost "unlimited" supply of interesting objects and slides and he prepared Wardian cases to act as table centerpieces.[38] Both parties were held at the Apothecaries' Hall at

Blackfriars and reflected all of Ward's passions—microscopes, Wardian cases, and medical science.

More than five hundred attended the first party, and at the second there were nearly seven hundred. The eminent guests included "almost every man of science in London."[39] Among them was Matthew Marshall, the chief cashier of the Bank of England, whose signature at the time was embossed on British banknotes. The evening afforded, as the *Illustrated London News* reported, "a treat of the highest and purest intellectual gratification."[40] Many of the distinguished guests thanked Ward personally for giving them a "bird's eye view" of the microscope. The following morning the Hall was visited by more than four hundred women and children, all equally keen to see the products of science. The size and grandeur of these parties should not be understated. In total as many as sixteen hundred people attended the events.

As Master of the Society of Apothecaries, Ward used his position to promote not only the society but also his interests, among them the Wardian case. "I was much gratified at the success of these parties," wrote Ward, "as all my friends rallied around me and gave me their warmest support."[41] The parties that Ward put on were well remembered by many scientists who attended, several of them giving it special reference when they reflected years later on Ward's achievements. Although nothing could equal the heights of the parties, following his tenure as Master he continued to work with the Society for the next decade in various capacities as treasurer and on the garden committee.

Throughout the 1850s the Society was increasingly buried in debt. One way out was to abandon the botanical garden at Chelsea. Added to this, a new embankment was being planned for the Thames that would all but remove the site. Ward led a group advocating to keep the garden, which had always served as an important training ground for young physicians. Ward not only showed its importance but led a project of renewal: new greenhouses, new layouts, and Wardian cases with exotic plants as showpieces throughout the garden. Together Ward and the curator, Thomas Moore, set about increasing the stock of plants by exchanging their duplicates with other gardens around the world. The

Chelsea Physic Garden has seen many moments of near abandonment over its 350-year history; on this occasion in the 1860s Ward played an important role in its survival.

But times were slowly changing. Well-connected amateurs' ability to effect change was slowly giving way to directions by professionals and bureaucrats. Furthermore, during this time Ward's health was failing. He suffered numerous illnesses, from an infection of the hand to pneumonia. Botany was often the pursuit that kept him going. In one period he could not leave The Ferns for three months. For entertainment he arranged the alpine plants that Joseph Hooker had sent him from Sikkam and Tibet according to elevation. "I think I should soon break down," he wrote on another occasion, "were it not for the recreations which botany affords."[42]

Death of an Amateur

In the 1860s Ward continued his work at Chelsea and in his garden at Clapham. Because time was upon him, he resolved not to have any more plants sent to him. "I had made up my mind not to purchase any more, but the temptation was too great," he said about purchasing more North American ferns.[43] Like all collectors, he harbored an overwhelming desire for new plants, a desire to hold the exotic, to plant it, and to watch it grow. Although he maintained friendships with like-minded people in faraway places, he never had the opportunity to visit them. He always wanted to see India after all the things Wight had sent him. And Gray continuously asked him to visit Boston. The farthest trip Ward managed to take was to the Continent. But he contented himself with his garden at Clapham, spending time among his plants from around the world, knowing their stories and their collectors, knowing the distance they had traveled toward him: "My great delight however is my little garden . . . reminding me of many kind friends."[44] He surrounded himself with the rhododendrons Joseph Hooker sent to him from India, the ferns Gray sent perfectly packed in cases, the arctic plants a friend sent him from Norway, the algae von Mueller sent him from Melbourne, and the cycads Stanger sent him from the Cape Colony.

In his last years, Ward grew increasingly disgruntled. He had already outlived his wife and many of his children. His daughter-in-law died and left his son with a three-month-old daughter. Another son, whom he had encouraged into the medical profession, was overwhelmed in his duties at the London hospital and suffered a breakdown. Finances also weighed upon him. For all his tireless work, he had little to leave his children. His investments before retiring did not yield their proposed returns, and he was almost destitute. In 1865 friends who had achieved far greater fame than him—William Hooker, John Lindley, and Joseph Paxton—all passed away.

In the end, his garden was his last pleasure. The Wardian case that he had so vigorously promoted and given to the public was a particular disappointment to him. His reflections were penned on Christmas Day of 1866 and are worth quoting in full:

> Thirty three years have elapsed since my first cases arrived in New Holland. You know what has been effected by them since. I have never received the slightest acknowledgement or thanks from any *public body* in this country. Independently of having had hundreds, I might say thousands of letters of enquiry to answer and all my leisure time and more than all occupied in receiving visits—from in too many cases— idle and ignorant people who were tired of their lives for want of something to do. But were my time to come over again, I should do precisely as I have done considering that my life, though one of constant labour, has been one of great delight.[45]

For three decades he had entertained people at his house, and on this gloomy Christmas, Ward wondered: to what end? As early as 1839 he had complained to William Hooker that much of his leisure time was taken up with answering questions regarding the case or receiving visitors at his house. Ward also turned his tirade to the government. Although disgruntled, there is clarity in old age and approaching death. And it is with this clarity Ward wrote to another of his close correspondents: "Had the Government taken as much interest in the welfare of the labouring populations, as in the introduction of tea/cinchona &c. into India we should not have had a such a report from the Commons."[46] He was referring to the report of the Commons that showed the harsh conditions in which the working class lived. Ward raised a good point. These transfers of tea and cinchona, which used his cases, appeared to have little benefitted the people of the working class, whom he had spent much of his life treating as a doctor. The plantations in distant locations were only benefiting colonists, the wealthy, and the economy.

Ward's outlook was tempered by his passion for plants. He was self-promoting, an obsessive collector, devout, and often liked to complain. But he was also a decent man who did a lot of good. He championed female doctors and also trained female gardeners at Chelsea.[47] He encouraged young people, whether Asa Gray or the children of his many friends. And above all he believed that his technology could benefit everyone and chose not to restrict it with patents.

There are many personally signed copies of Ward's books *On the Growth of Plants in Closely Glazed Cases* deposited all around the globe. There is one signed to Ferdinand von Mueller at the Royal Botanic Gardens in Melbourne, another at the Gray Herbarium at Harvard, and yet another at Kew Gardens. But maybe the notations in Ward's personal copy speak volumes here. He inscribed his copy with a quote from the old *Spectator*: "The consciousness of approving oneself a benefactor to mankind is the noblest recompense for being so."[48] There was the argument against glass tax, and the prescription of Wardian cases for the poor, and the insistence on the possibilities of his case for betterment. All of this was wrapped up in his piousness, but it still showed a commitment to use his invention to help others.

Ward nonetheless remained accepting of his fate. His love of natural history had brought him many riches. He wrote to Asa Gray in Boston, "The love of natural science and the acquisition of wealth are ... incompatible."[49] Ward went on, "Do not think that I regret the election I have made, as all the riches in the world could not have afforded me ... more." He closed out his days among his plants and among the memories of his friends.

Nathaniel Bagshaw Ward died at St. Leonards-on-Sea on 4 June 1868. He was survived by two daughters and a son. When the *Morning*

Post reported his death, the headline said simply, "Inventor of Wardian Cases."[50] The short obituary captured his career in only a few comments: the Wardian cases, his connection with the Society of Apothecaries, his mentoring of young men and women into medical practice and natural science, and finally his house—"one of the scientific curiosities of London, showing how many and how well plants might be cultivated in a small space in or near a large city." There was an outpouring of obituaries and praise by his friends for the Wardian case and his tireless efforts. Joseph Hooker described him "the most useful and pleasant" correspondent he ever had, going on to say that "a large proportion of the most valuable economic and other tropical plants now cultivated in England would but for these cases, not yet have been introduced."[51]

But a prickly question lurks in the background. Many other inventions do not bear the name of their inventor, so why is it that Ward's name was attached to these boxes?

The Wardian case was not just a box for moving plants; it was the name of a network. No one illustrates this point better than Nathaniel Ward himself. He was certainly the catalyst for the closed system of plant boxes. But he was not a big publisher: he published only one major work (and republished it a decade later). We even know that the basic construction of the Wardian case was available in the late eighteenth century. Instead, it was above all Ward's far-flung friendships and influence that secured his fame. Although Ward died destitute, bemoaning the effort he had made toward showing people his cases, his name has been preserved for posterity—attached intimately to the cases. Science is like that; it remembers those at the center of the network. Most of the time they are men, most of the time they are white, and most of the time they are of European descent. But rarely are they amateurs with as few means as Ward. At a time when most scientists had large botanical gardens and colonial powers to help their cause, Ward had his curious house.

The death of the inventor, however, is not the death of the technology. Had the story of the Wardian ended with Ward, its impact and range would warrant a far less dramatic story. The utility and widespread use of the case carried on for more than half a century after Ward's death. It is to those stories we now turn.

In the late nineteenth century, two major shifts show that the time when a simple gardener living in Clapham could wield such wide influence was coming to an end. First, the rapid expansion of the commercial nursery trade increased, both in variety and in volume, the demand for exotic plants. Following the Great Exhibition, gardening increased dramatically among the middle classes. With such large commercial interests involved, nurseries became ever more important. Second, the institutionalization of economic plants, especially toward the ends of imperialism and environmental management more generally, demanded professional scientists serving the needs of empire. Nonetheless, even though the amateur's role slowly died, the Wardian case still had an important role to play for the next half century.

PART 2

Panoramas

7

Logistics of Beauty

Only two years after the death of Nathaniel Ward, the shrewd nursery-man William Bull took out a patent for "An Improved Case for the Conveyance of Plants." In fact, it was a Wardian case with a few changes—louvers to block the sun and air holes for circulation. By 1870 Bull's nursery had garnered a reputation as an excellent supplier of exotic plants. He had been in the business less than a decade but already had a long list of subscribers. He also had a flourishing establishment on King's Road, Chelsea, at the heart of London's thriving nursery trading district. After Ward withered away almost destitute, Bull offered his improved case to be used by others if they paid a "moderate royalty."[1]

The commercial nursery trade was made up of the innovators, brokers, and movers of plants, and the Wardian case was an important tool in their business. In the 1860s and 1870s we see that it was nursery firms that played an important role in innovating the case. William Bull was one of many important firms that moved plants in Wardian cases. The use of the case was not limited to British nurseries. In Europe there were firms such as Godefroy-Lebeuf in France and Haage & Schmidt in Germany; and in Australia there were firms like Thomas Lang. All over the world there were nursery firms that had a commercial interest in the successful movement of plants. Gardening was a wildly popular pastime in the nineteenth century, and agriculture was a spreading pursuit of colonizers. For both of these groups, nurserymen were the suppliers.

How commercial nursery firms used the Wardian case is an important theme in the history of the plant case. Not only did they use it, but the volume of their dispersals was immense. The commercial nursery trade was global and extensive, and the Wardian case helped move their plants; and as we will see, in the twentieth century these had major environmental consequences. But first let's consider the beauty they traded in, the plant hunters they sent out, and the enormous trade they conducted.

Bull's Improved Case

William Bull was one of the leaders in the exotic nursery trade and one of the last of the great nurserymen of Chelsea.[2] He sourced plants from Africa, India, Japan, and North America. He supplied to the queen and the prince of Wales; to royalty in Austria, France, Germany, Russia, and Sweden; and to "most members of the aristocracy in the United Kingdom interested in horticulture."[3] In 1863, after years as a plant hunter for other firms, he commenced his own operation by setting up shop in the fashionable neighborhood of Chelsea. In the early days he focused on greenhouse plants, specializing in fuchsias, verbenas, and pelargoniums. As for the last-named, he is remembered for introducing the delightful "Chelsea Gem," which has rounded leaves ringed with creamy white and pink double flowers. He was also an astute businessman, sourcing the rarest and choicest plants, instigating crazes, and influencing styles.[4] Soon he turned to orchids and, along with Veitch nurseries, was at the forefront of the orchid craze that gripped the gardening world in the late nineteenth century.

Bull's plant case was an innovation decades in the making. Horticulturalists, nurserymen, and plant hunters had tinkered and experimented with designs for the case, but no one design emerged. Bull capitalized on all of these practice-based innovations in taking out a patent for his plant case (fig. 7.1). When his plans were released to the public, the *Gardeners' Chronicle* was quick to point out that while Ward's case was a "great step" in allowing nurserymen to move novelties all over the globe, it was

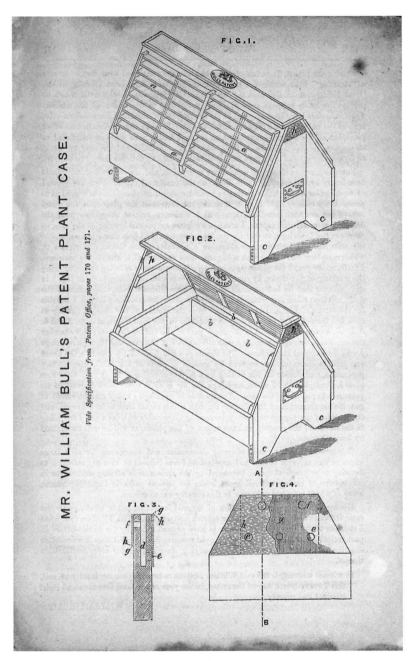

FIG.1.

FIG.2.

FIG.3.

FIG.4.

MR. WILLIAM BULL'S PATENT PLANT CASE.

Vide Specification from Patent Office, pages 170 and 171.

FIGURE 7.1 William Bull's "Improved Case for the Conveyance of Plants," patented in 1870. From William Bull, *A Wholesale List of New and Beautiful Plants* (London, 1871).

not faultless. Often, as we have seen, cases arrived at their destination filled with dead plants. For botanical gardens this was undesirable, but for commercial nursery firms it meant catastrophic financial losses. Bull's improved case was the result of "great observation and experience." As a large importer and exporter of plants, he saw two chief problems with the Wardian case. The first was the "direct action of the rays of the sun upon the glass," which caused the plants to steam to death. Second, was the "frequent" breaking of the glass, which left plants exposed to the harmful effects of the sea air.[5]

Bull devised a number of solutions. Many of these changes, we might recall, were already known in the eighteenth century. First, he added horizontal louvers as a protective frame on top of the glass inserts on the sloping roof of the case. This not only filtered the sun but strengthened the glass. Second, inside the case, underneath the sloping cover, he added internal channels filled with sphagnum moss or charcoal to absorb moisture, thus allowing the condensation coming off the glass to be collected. Third, he added feet to the bottom of the box to lift the cases off the deck so that seawater did not enter them when the decks were washed. Fourth, one-inch air holes covered with fine wire mesh were added on either side of the box to allow the "escape of foul and overheated air from the interior of the case."[6]

The new case, however, was based upon ideas developed by many other horticulturalists over a span of decades. The Wardian case was a practice-based technology that evolved over many years thanks to tweaks and innovations by many different makers and users. The louvers were a new, unique addition and were to remain an important part of the case's construction for the next half century. The small air holes represented a recognition of the importance of circulating air—a fact widely known in the eighteenth century. The channels for collecting condensation would become unnecessary because the small air holes limited the condensation affect. Putting feet on the case had been widely known ever since Robert Fortune utilized the cases in the 1840s. Nursery firms and institutions were quick to adopt the new designs. Places such as Kew, the Jardin des Plantes, and the United States Department of Agriculture were to use a similar case for many years to come. While Bull was the leading promoter for modernizing the design of the plant case, his name would never be attached to it. The boxes would remain "Wardian cases" in the common vernacular.

Bull supplied to his customers both exotic garden plants and important economic varieties, such as coffee, grapevines, and fruit trees. Following the collapse of the coffee industry in Ceylon and India in 1869 from disease, the astute Bull brokered a solution. In 1872 he received his first shipment of Liberian coffee from West Africa, the same year that Kew received their first shipment. Bull labeled the variety *Coffea liberica*, advertised it in his catalog, and sold it to desperate planters in the colonies.[7] Liberian coffee was much hardier and more resilient to the fungus devastating the existing plantations. By 1880 Bull had sent his newly patented cases full of the new variety of coffee to plantations in Ceylon, India, Java, Brazil, and Central America. According to one report, "tens of thousands of sturdy seedlings were shipped from the Chelsea establishment to almost every coffee growing establishment."[8] Nursery firms were always very important in moving large numbers of economically valuable plants to satisfy the needs of planters.

In the late nineteenth century commercial nurseries were at the forefront of the global transport of plants. Their repeated shipments made them acutely aware of the needs of live plants on long journeys, and their livelihood depended on reliability. The success of a plant case was critical to their trade. They had plant hunters in the distant parts of the globe, extensive areas of greenhouses set aside for the propagation of new plants, and numerous patrons to whom they could distribute new exotic plants. In their seasonal catalogs they had sections of "New Plants Offered for the First Time." And it was here that patrons could read of plants traveling in Wardian cases.[9] The taste for plants was not limited to the upper class; but the emerging middle classes were turning enthusiastically to gardening. And the spread of empires saw the rise of savvy gardeners all over the globe.

Royal Exotic Nursery, Chelsea

Bull's nursery was located at 536 King's Road, Chelsea. By the time he took up premises on King's Road, it was the hub of nursery activity in

London: more than twenty-five nursery firms operated from storefronts along the road. Here were all the major exotic nurseries, among them Christopher Gray's Nursery and Colvill of King's Road.[10] Many firms operated out of different locations around the country but rented a shop front on King's Road to maintain their reputation. It was also a short walk to the Chelsea Physic Garden that Ward and Moore had worked so tirelessly to renew. Indeed, Moore was now editor of the *Gardeners' Chronicle*, the leading gardening magazine of the period, and was in an ideal location for seeing the newest and choicest exotic plants arrive in the city. In the nineteenth century greenhouse and conservatory plants were the flavor of the day. Hardy plants suitable to the climate would not come into their own until after 1914.

The most famous nursery on King's Road was the Royal Exotic Nursery of James Veitch & Sons.[11] Originally from Exeter, they set up shop at 544 King's Road in 1852, following the Great Exhibition. They quickly became the biggest and best. Importantly, this was at the same time that Loddiges's nursery at Hackney was closing its doors and putting their extensive collection under the gavel. Many of Loddiges's plants, particularly orchids, were purchased by Veitch & Sons and taken to their new establishment at Chelsea.

Veitch & Sons was a horticultural dynasty.[12] Three important plantsmen headed up the business during the nineteenth century: James Veitch Sr., who brought the business to prominence from rural Exeter with his skill as a gardener, his ability as a scientist, and his business acumen; his son, James Veitch Jr., who inherited many of his father's successful attributes and took them to Chelsea to successfully establish the Royal Exotic Nursery; and Harry Veitch, who in the 1870s took over the Chelsea business following the death of his father and elder brother and shepherded the business into the early twentieth century. Harry expanded the business dramatically and was deeply involved with the Royal Horticultural Society. One of Harry's lasting achievements was to oversee the first Chelsea Flower Show. All three Veitches used the Wardian case to great effect in conducting their business.

The Wardian case was first used by the Veitch nursery when James Sr. sent out their first plant hunters, William and Thomas Lobb. William

Lobb was the first brother to leave. In 1842 at the suggestion of William Hooker, he went to South America and followed some of the travels of George Gardner. Lobb then crossed the Andes and while in Ecuador had Wardian cases made on-site by local carpenters. After three years abroad he returned to Britain with many rare plants packed in Wardian cases.[13]

Following his brother, Thomas Lobb was sent to Southeast Asia and was equally successful in finding rare plants. Like his brother, he found sending them back to be a challenge. One of his consignments arrived back at the nursery in the middle of a freezing winter of 1844 and all the plants were dead. Veitch lamented in a letter to William Hooker, "So disappointing to have things brought safely almost to your door and then lost for want of care."[14] But there was more luck in the next consignment. One of the prized plants unpacked from the Wardian cases was the orchid *Cypripedium barbatum*.[15] Over the next two decades the Lobb brothers collected plants from all over the globe. Many of their prized plants arrived back in Britain in Wardian cases.

The process of sending out plant hunters brought the Veitches into close contact with scientists. Many of their collectors, such as the Lobb brothers, prepared herbarium specimens for William Hooker. In return Hooker gave Veitch valuable information on locations where plant hunters should collect, named their new plants, and promoted them in well-illustrated publications such as *Curtis's Botanical Magazine*. Often a plant was not named for sale until it had appeared, in all its color-plate glory, in the pages of a respected magazine.[16]

There were also other relationships that benefited science, empire, and commercial nurseries. Indigenous people in Southeast Asia demonstrated the value of the natural latex substance gutta percha, but few Europeans knew what plant it came from. Thomas Lobb, while collecting for Veitch, followed advice from native Malayan people in and around present-day Singapore, collected the plant, and sent it to Kew. William Hooker named it *Isonandra gutta* (today *Palaquium gutta*). Following the discovery of gutta percha, the plants were widely circulated using Wardian cases. The substance was a vital commercial product in the late nineteenth century: among its many valuable uses was coating the underwater telegraph wires that were spreading across the world. But collection of the plant also led to the destruction of many forests around Singapore.[17] After the Lobb brothers, there were many other Veitch collectors, all of whom used the Wardian case to bring into cultivation rare and desirable plants.

"Fair Winds and Smooth Seas" for Japanese Plants

Following Japan's opening itself up to trade after the Treaty of Kanagawa, plant hunters working for nursery firms soon arrived to find the newest exotic plants. John Gould Veitch, the son of James Jr., traveled there in 1860. His ship was wrecked off the coast of Sri Lanka, and he lost all his possessions, including the Wardian cases he had brought with him. He eventually made it to Japan. Although John Gould undertook a number of adventurous expeditions, including one to the top of Mount Fuji, many of the plants he found were from local nurseries, often collected through the help of local guides. In Nagasaki he set up a garden to care for all the plants he was collecting. From there they would go into cases to be shipped back to Chelsea.

John Gould contracted Japanese carpenters to make Wardian cases for him. They found the concept of the cases a strange one; as Veitch wrote in a letter to Chelsea: "The glass cases quite bewilder them; I had some trouble to make the carpenter believe I was in earnest ordering them. They think me mad to try and send plants to England in this manner."[18] In spite of their doubts, the Japanese built many Wardian cases for him. At his garden he found a "capital place" to store the cases: he transferred plants into the cases and erected a bamboo shed to protect them from the sun.[19] The next step was to wait for ships heading to Hong Kong or Shanghai that could transport the cases on the first leg of the journey.

Robert Fortune, after his travels in China, was in Japan at the same time as John Gould in search of plants for the nursery of John Standish of Bagshot. Fortune had his cases made on-site in Yokohama, but the Japanese carpenters would not do the glazing for him, so he had a Dutch carpenter insert the glass panels. As Fortune noted, "In a foreign country . . . even Ward's cases cannot be made without some difficulty."[20] Fortune traveled widely in Japan and found many valuable plants. He left many of his plants with the American trader George Rogers Hall, who had a garden in Yokohama.

Fortune and Veitch did not cross paths in Japan, but as Fortune was preparing to leave, he stepped aboard the SS *England* and found the poop deck almost full of Wardian cases—Veitch's cases. Fortune added his cases to the poop deck and hoped for a safe passage for the plants. He later wrote in his book *Yedo and Peking* (1863): "Never before had such an interesting and valuable collection of plants occupied the deck of any vessel, and most devoutly did we hope that our beloved plants might be favoured with fair winds and smooth seas, and with as little salt water as possible."[21] Their plants arrived in very good health, and three days later many of Fortune's plants were shown at the London Horticultural Society.

John Gould sent many plants and seeds back to England. By the time he left Japan he had already sent at least thirteen glass cases. Most surprising is that his letters home, which were edited before being published in the *Gardeners' Chronicle*, reveal that half of those cases were sent to institutions and dignitaries as gifts. He sent four cases to Kew and one to the French minister who had helped him in Japan. Another, filled with tea seedlings, went to Bombay for the captain of the HMS *Berenice*. And he sent another to the governor of Hong Kong to thank him for his help in setting up the new botanical gardens. He wrote of the gifts: "These little matters take up much time; but having received many kindnesses, I am anxious and glad to show my gratitude as far as I possibly can."[22] Many gifts and favors were needed to smooth the politics of plant collecting.

After nearly two years in Japan, Veitch sent back six more Wardian cases to Chelsea. Some of the most well-known Japanese introductions that the Veitches brought into circulation are *M. liliiflora* and *M. stellata*, and probably one of the most common of all Japanese plants sent back by Veitch was the Japanese ivy.[23] One of the most widely desired introductions was the umbrella pine, *Sciadopitys verticillata*. There was some controversy as to whether it was Veitch or Fortune who introduced the plant,[24] but regardless of their competing claims both had most likely gotten them from local Japanese nurseries near Yokohama. A few years later John Gould set sail again, this time to the South Pacific, and again he used the Wardian case to send back plants to the Exotic Nursery in Chelsea.

While Fortune shared the ship's deck with Veitch, he shared a garden in Yokohama with George Rogers Hall. Originally trained as a doctor, Hall turned to trading in curios, at which he made a substantial profit, after journeying through China and Japan; but his passion was plants. He gathered many Japanese plants in his garden in Yokohama. As Fortune traveled around the country, he let his plants grow in Hall's garden. When Fortune had Wardian cases made for sending plants back to England, Hall had cases made to send to the United States. While Fortune's plants were destined for the Standish nursery, Hall shipped his plants back with the hope of selling his plants to nurseries in New York.

In March 1862 Hall wandered into the offices of the well-known New York exotic nursery Parsons & Co. in Flushing and offered his collection for sale. After a long conversation, Hall and Parsons agreed on a sum for the Japanese plants. The following day Hall walked into their offices with a large collection of plants packed in Wardian cases. There was much excitement in the New York offices awaiting such a rare collection of plants. Parsons later wrote of the excitement in the *Horticulturalist*, the leading American horticulture journal of the day: "If you have ever seen the eagerness with which a connoisseur in pictures superintends the unpacking of some gems of art, among which he thinks he may possibly find an original Raphael or Murillo, you will have some idea of the interest with which all, both employers and propagators, surrounded those cases while they were being opened."[25] There were many novelties, a number of them never before seen on American shores. As the editors of the *Horticulturalist* noted: "So many fine things have never before been introduced at one time. Our English brethren pride themselves, and justly, on their enterprise in collecting rare and beautiful plants from all parts of the world; we have done comparatively little."

Hall's was a diverse and beautiful collection. The 1862 consignment for Parsons included chrysanthemums, maples, Japanese barberry, magnolias, bamboo, columbines, oaks, wisterias, and many more. There were also Japanese chestnuts, surely some of the first to arrive in New York.

The previous year Hall had sent another consignment of plants to Francis Parkman's small three-acre summer estate at Jamaica Pond in Boston, not far from the location of the Arnold Arboretum.[26] Packed in these Wardian cases were umbrella pines, dogwoods, rhododendrons, crabapples, and cypress pines. They all survived the seventy-day journey from Yokohama to Boston.[27]

After the first purchase, Parsons also bought another six Wardian cases from Hall. When they arrived in New York, Parsons found among many other plants a new variety of Japanese honeysuckle (*Lonicera japonica*). It became very popular among gardeners on both sides of the Atlantic for its fragrant flowers.[28] The editor of the *Horticulturalist* concluded enthusiastically on Hall's large collection: "Let us hope that the splendid collection now placed in the hands of the Messrs. Parsons will, in this respect, mark a new era in our history."[29]

Many plants that arrived in Wardian cases have swept across North America. Not only have these plants, such as the umbrella pine, become a beautiful addition to landscapes and gardens, but some, such as the Japanese honeysuckle, have spread throughout woodlands across the eastern United States; it is one of the most invasive species in North America. Indeed, even in Hall's own garden in Bristol, Rhode Island, where he planted a small selection of his Japanese plants, by the 1920s the honeysuckle had "run rampant" and "done much harm" by choking the conifers.[30] Early varieties of the honeysuckle had arrived at Kew as early as 1806, but ones like those Hall collected after the opening up of Japan spread rapidly with the nursery trade, not only in North America but also across Britain and Europe. The Japanese honeysuckle has become established every continent on earth except Antarctica. It is but one example of the many invasive plants that have been spread in the Wardian case as a result of the nursery trade.

Beauty in the Antipodes

From the 1850s on ferns were all the rage in Victorian Britain. This fern craze, as it was called, is closely connected to Ward's invention. Ferns displayed under glass in a parlor-style Wardian case epitomized refinement and as an indoor adornment completed the most tasteful living space. The fern craze would not have taken off, or lasted so long, if the parlor cases had not become readily available. As the craze developed, more and more people demanded exotic varieties. Often people went to nurseries for new and rare specimens. One of the most famous fern suppliers of the time was James Backhouse & Sons of York. They were widely credited as Britain's leading cultivators of exotic ferns.[31]

At Backhouse's York nursery a number of greenhouses were set out as display rooms that re-created tropical and distant environments in which customers could experience exotic landscapes. One of their greenhouses, over seventy feet long, was complete with an Australian–New Zealand fernery. One *Gardeners' Chronicle* reporter described it as "a fascinating imitation of wild Nature . . . the whole combining to form a sight which it would be difficult to persuade a Maori had not been bodily transferred from the Antipodes."[32] These ferneries were re-creations of dense green scenes from distant parts of the globe; they were lauded at the time, but upon reflection they also display the artificial manipulation of nature that characterized the Victorian era.[33]

For such a scene of ferns to be created and then sold to customers, the plants had to travel. There were many rare ferns from the Antipodes in Backhouse's display rooms. Of wide interest were those that it might be possible to grow outdoors in Britain. The Backhouses were in the process of acclimatizing the *Gleichenia* genus of ferns to be hardy enough for the British climate. One suitable but rare variety was *Gleichenia alpina*, native to Tasmania.

As for all nurseries trying to acquire rare plants from around the globe, contacts were important to Backhouse's trade. Outfitting a plant explorer was an expensive task, so paying a colonist to acquire plants often seemed to make sense. But there were costs to this approach too. A Wardian case of Tasmanian ferns sent to York was overwatered before being shipped, roughly handled by the sailors, and in transit for over nine months, so they arrived in poor condition. All of the *G. alpina* that Backhouse had so desired were dead. In 1860 the nursery asked the enthusiastic young lawyer James Walker, a close family friend, to acquire new specimens of *G. alpina*. They were very specific on the collecting location: they wanted them *only* from the top of Mount Wellington—the grand snow-capped peak that overlooks Hobart.[34]

Getting Wardian cases out to the distant colonists presented still more difficulties. James Backhouse Jr. wrote to Walker requesting the collector to have two Wardian cases made on-site in Hobart.[35] What followed in the correspondence between the nurseryman and the young colonist was a detailed description of how to construct and pack the cases, which included a sketch of the case to be built. Backhouse, like many nurserymen, was very concerned about moisture buildup in the case. Inexperienced gardeners had a tendency to overwater plants. Backhouse recommended drilling two holes in the bottom of the box for drainage. Once packed, the cases should be left open for at least twelve hours to let excess moisture content evaporate. In addition, ebony iron handles were be to be affixed on each side so that the case could be fastened to the deck. The cases were to be made of one-inch-thick wood. And the thick plate glass was to be covered with wire gauze to protect it from breakage. The plants inside were to be packed close together (this was as much about economy as about safety). Backhouse not only paid for the cost of constructing the cases and for the expenses incurred in getting the plants, but also paid Walker £5 for the trouble. Interestingly, he was also prepared to pay up to twenty shillings to the ship captain taking the cases if he promised to care for the plants. The effort paid off. The two cases arrived in York in the summer of 1861, and although the ferns were not thriving, they would return to health.[36]

For the large nurseries in Britain and Europe, getting the plants they needed often meant being closely connected to the colonies. It also meant informing those collectors of the best means of transporting plants—how they should construct and pack the cases. The fern craze was a long-lasting botanical passion from which nursery firms profited. The Backhouse nursery is representative of other large nurseries that specialized in offering ferns to keen local gardeners, such as Stansfield in Lancashire, James Ivery & Son in Surrey, Lee & Kennedy in Hammersmith, Low & Sons of Clapton Nursery, and William Rollison & Sons of Tooting, as well as William Bull and Veitch & Sons.[37]

For the social historian David Allen, the Victorian fern craze was possibly "the greatest and ultimately most destructive natural history fashion of all."[38] In Britain people ravaged forests to get ferns. But forests all over the globe were trampled in the quest to find new ferns to send back to nurseries in Britain and around the world.

In the nineteenth century, moving plants was a global industry. There were also nursery firms in Australia and other faraway places that were intimately linked to the major exotic nurseries in London, such as Veitch and William Bull. Nursery firms expanded dramatically, their global reach and contacts proving that very specific species from the most remote regions could be requested and delivered effectively. But the trade went in both ways. Nurseries in places like Australia could request choice exotic plants from Britain to be sold in the colonies.

One of the most successful nurserymen in 1860s Australia was Thomas Lang in Victoria. Lang set up his nursery in Ballarat in 1856. Over the coming years he went from selling such common herbs as thyme and sage to importing showy exotics.[39] At the time the small, rural Australian town of Ballarat was undergoing dramatic economic and social change following the discovery of gold. It had a number of societies, including one devoted to horticulture.

In early April 1862, Lang delivered a lecture to the Ballarat Horticultural Society titled "On Wardian, or Plant Cases."[40] He began by describing his first encounter with the cases in Edinburgh, in the house of James McNab. The plants in those cases, Lang was surprised to learn, had gone twelve months without watering. The lecture was a lengthy argument for the "usefulness" of the cases for horticulturalists, illustrated by the story of Robert Fortune and tea. But the best example was the plants that Lang had himself introduced, such as the giant Californian redwood that he brought to Australia with the help of the case. He brought with him two Wardian cases, just arrived in Ballarat from England, that carried "18 of the newest and finest varieties of fuchsias to be had in England." The cases also held camellias, ferns, and orchids.

To see the Wardian case firsthand allowed the people in the room to materially grasp the work being done to transform the colony into a place of usefulness and beauty. But colonists also envisaged enthusiastically "the comfort, the pleasures, the commercial interests, the happiness of mankind are promoted by the use of Wardian cases," as Lang proclaimed.[41] A colonial nursery was one of the many practices that were significant in taming a foreign land.

In the 1860s Lang imported hundreds of Wardian cases from well-known nurseries in Britain, among them: Veitch & Sons in Chelsea, Thomas Rivers in Hertfordshire, and Hugh Low in Clapton. Not only did he import plants, but he propagated them at his nursery and sent out plants to buyers in Australia, New Zealand, the Pacific Islands, and India.[42] In 1870, after just over a decade in business, Lang told his customers, "We have brought hither very nearly ONE MILLION living Plants."[43] One million plants—it was a huge number for a single colonial nursery.

Consuming Plants

Desire was at the heart of the nursery trade. The desire to know the botanical riches of the world, the desire to showcase the exotic, the desire to possess the exotic. Of all the stories that display the deep desire that drove many parts of this trade, the story of the pitcher plant, *Nepenthes northiana*, is one of the most illustrative. It began not with a plant hunter but with a British artist. The pitcher plant is carnivorous, meaning that it consumes insects and even rodents to survive. The prey is ensnared in the large pitfall trap—the pitcher—and through chemical processes they are turned into nitrogen and phosphorous to feed the plant. *N. northiana* is one of the largest pitcher plants known to exist: its traps can grow up to a foot in length. There was a great desire for these unique plants, and many of the largest varieties were moved in Wardian cases.[44]

Marianne North was a restless and intrepid British traveling botanical artist. Between 1871 and 1885 she went to fourteen countries, all the while painting local plants and landscapes.[45] One of the lasting wonders of North's work was the production of images that went against the grain of the formal botanical illustrations of the period. Many of her paintings were set in the wild, giving viewers back home an appreciation of how plants grew in their environment as opposed to the greenhouse.

In 1876, while traveling through Java and Borneo, North ventured deep into the jungle near Sarawak.[46] One night she stayed at an abandoned antimony mine while her guide, Herbert Everitt, an employee of the Borneo Company, went farther into the jungle to bring back specimens of large pitcher plants that grew in the area. "When I received them," wrote North, "I tied them in festoons all round the verandah, and grumbled at having only one small half-sheet of paper left to paint them on."[47] The result was a wonderful painting. Dangling green and red traps marked with their various splatters of red and purple were figured in the foreground of the Limestone Mountains that overlook Sarawak.

The small painting completed by North could not do justice to the large plants. But it still garnered much interest when she showed it to nurserymen in London. Harry Veitch, who was by now in charge of the Royal Exotic Nursery, was taken with it and soon afterward sent out the plant hunter Charles Curtis, who found the pitcher plant and sent it back to Chelsea in Wardian cases. Specimens were sent to Kew for naming, and Joseph Hooker named it *Nepenthes northiana* (fig. 7.2).[48]

Desire drove the demand for many exotic plants. Plants evolve to meet the needs of their pollinators. One orchid (*Bulbophyllum beccarii*) found by a Veitch collector smelled of rotting fish in order to attract its chief pollinators, carrion flies. In *N. northiana* Veitch saw an interesting and unique plant and decided to acquire it for his Chelsea nursery. That it had already been painted by a well-known artist made it even more saleable. Veitch wanted to attract customers desiring the rarest and most interesting plants, and a pitcher plant almost a foot long certainly fit the bill. Desire for something new was a driving force in all of these crazes and ventures. Today *N. northiana*, like many carnivorous plants, is a rare species in the wild. Owing to overcollecting from the commercial nursery trade, it is listed as vulnerable on the Red List by the International Union for the Conservation of Nature.[49]

Selling Plants, Trading Nature

The vast trade in exotic plants was not confined to British nurseries. There were companies such as Jean Linden's in Brussels, which specialized in orchids; Haage & Schmidt in Germany; and, by the 1890s,

FIGURE 7.2 Inside the *Nepenthes* (pitcher plant) house at the Royal Exotic Nursery, 1872. Many pitcher plants were moved from Borneo in Wardian cases. From *Gardeners' Chronicle*, 16 March 1872, 359.

Yokohama Nursery, among others, in Japan. As we saw in Melbourne, nurseries in the colonies also became important movers of plants, and nursery firms cropped up all over the globe (see fig. 7.3). They issued to interested customers catalogs listing an incredible variety of plants. That degree of diversity is unimaginable today: in Melbourne, heritage practitioners maintaining historic buildings and their gardens for the National Trust cannot now re-create the gardens of the nineteenth century be-

FIGURE 7.3 All over the world local workers played an important role in maintaining the supply of exotic plants for the nursery trade. Here workers in Ichang, China, pack lily bulbs to send back to the Farquhar nursery in Boston, ca. 1907. Photo by E. H. Wilson. © President and Fellows of Harvard College, Arnold Arboretum Archives.

cause many of the plants that populated them are not available.[50] These catalogs also offer insights into dispersal. Take for example the Japanese knotweed, one of the most invasive species in Britain. From 1869 to 1936 twenty-one nursery firms offered the plant for sale,[51] including those of Backhouse, Bull, Standish, and Veitch.

Nurseries introduced and spread many plants that are now a problem. The biologist Richard Mack has done a vast amount of work on the history of the nursery trade. After reviewing hundreds of nineteenth-century nursery catalogs, not for their ephemeral value but for their scientific content, he noted that the "results have been both beautiful and disastrous."[52] The live plant trade is one of the main ways in which plant pests are spread around the world. These pests have severe ecological and economic consequences.[53] In Europe it is estimated that 38 percent of invasive arthropods were imported by the horticultural and live plant trade. In the United Kingdom nearly 90 percent of invertebrate pests are believed to have been imported on live plants. In the United States nearly 70 percent of insects and pathogens that impacted forests between 1860 and 2006 traveled on imported live plants. History plays an important role in understanding these complex movements. Often there are long time lags between when a species is moved and when it is so well established as to be a problem.[54]

The early twentieth century saw dramatic changes in the British nursery trade. William Bull's son Edward took over the King's Road firm and continued to trade in orchids, surviving in the early twentieth century only by dropping the price. With no one to take over the family dynasty, Veitch closed its doors in late 1914, and the plants and equipment were sold at auction. A few years later Bull's nursery also closed its doors. It was one of the last nurseries to close on King's Road. The nursery trade, however, continued to gather momentum. In the early decades of the twentieth century, as shipping became faster, international wholesale nursery firms in places such as Holland, Japan, and Indonesia were able to capture more and more of the market.

8

Kew's Case

Wardian case no. 10 began its life as wood and glass and nails in the workshop at Kew Gardens. It was constructed in early 1858. Soon afterward it was packed with eleven valuable plants, including pears, gooseberries, oranges, plums, and oaks; of the last, there were three different varieties from Assam. Destined for a Dr. Sinclair in Auckland, the case left London in January 1858. Case no. 10 was accompanied by no. 12, which contained many beautiful ornamentals: roses, daphnes, cedars, rhododendrons, and birches. Seeds of roses were sown into the soil of the two cases. Of all the plants sent out from Kew, roses and fuchsias were among the most common. The cases arrived in New Zealand relatively healthy. They were unpacked, and the plants were distributed throughout the colony.[1]

On the last day of 1858, no. 10 returned home to London in good condition. It sat near the greenhouse at Kew for three months. In March 1859 it was readied for another shipment, this time for Jamaica.[2] In it were more ornamental plants—fuchsias, begonias, and ten varieties of roses—as well as the unique Abyssinian banana (*Musa ensete*, now *Ensete vitricosum*). It was sent to the amateur naturalist William Thomas March in Kingston. This time the round trip was shorter, about six months. On 6 October, case no. 10 returned home for the second time.

March had packed it with nineteen varieties of Jamaican plants, including an unnamed prickly yellow shrub, a greenheart, and a red beefwood. While the plants that had gone out were thoroughly described and well-known, the ones returning went by many names, often local ones bestowed by colonists and local workers. The curators and botanists at Kew propagated the plants, closely assessed the plants, and fitted them into their systems of classification.

Case no. 10 did not sit around for long: by 26 October it had been repacked and sent on its way again, this time to Mauritius. Interestingly, this time the case was not filled with soil but coconut fiber. It contained forty-three different plants, including fourteen varieties of the Mexican species *achimenes* and six varieties of the elephant ear plant (*Caladium* spp.). The plants arrived at the botanical garden for the curator, James Duncan, to distribute. It sat in Mauritius nearly a whole year, finally returning to Kew in September 1860 packed with twenty-one varieties, many of them valuable—for example, *Hugonia serrata*, a "native climber" with fine orange flowers, and an unnamed plant with beautiful white flowers "from the top of one of our mountains"—but all "in a poor state."[3] It was accompanied by case no. 24, which had been on its own worldwide journeys before arriving in Mauritius, although this one had sat at the botanical garden for nearly *two* years before making the return journey.

Case No. 10 was not used again. But in its brief two-year journeys it nearly circumnavigated the globe and moved many plants. Some of these plants were useful, some were beautiful, but there was always a great variety. The Wardian cases employed by Kew moved thousands, probably hundreds of thousands, of plants around the globe. In the late 1850s there was a brief period during which the curators in the greenhouses numbered the Wardian cases being sent out; no. 10 was one of these. The numbering went as high as no. 41 (which went to Gambia and back). The practice of numbering the cases only lasted a decade, but it gives a glimpse of how these boxes were sent on multiple journeys and with many different plants.

Kew Gardens has a long and storied history as one of the world's leading botanical institutions.[4] It is widely recognized that the Wardian case was an important technology for the gardens to employ. As the historian Lynn Barber notes, "Kew's Wardian cases were probably one of the best investments the British Government has ever made."[5] But how was the Wardian case used, and how did the plant trade operate?

This chapter proceeds in two parts. First, it analyzes the period from 1842 to 1865, when William Hooker was director of Kew, to understand the movement of Wardian cases into and out of Kew. Second, it charts the importance of the Wardian case in transplanting rubber (*Hevea brasiliensis*) plants. The latter was probably the most important transplant the Wardian case ever facilitated.

The Plant Books, 1842–1865

The Wardian case was indiscriminate in moving plants; it carried whatever was planted. Plants arriving in Wardian cases satisfied both botanists in imperial centers and the floristic ambitions of settler societies. To follow the Wardian case is to follow a fragmented historical record—fleeting notations on how plants were moved. Botanical gardens commonly recorded details regarding the transport of plants themselves—the plant, the shipper and/or receiver, and the place—but rarely the means of transport.

The reason for this is quite simple: utilization of the Wardian case was so widespread that it did not seem remarkable in any way. Furthermore, once a certain manner of record keeping began, it was followed for decades. But there is one location where the movements of Wardian cases were recorded consistently—Kew Gardens. Over a period of two centuries the Goods Inwards/Outwards Books, as they are called, recorded the comings and goings of Kew's plants (fig. 8.1). In 1838 they show the first four Wardian cases being sent from Kew to the new colony in Western Australia (see chapter 3). From that point on the greenhouse keepers kept detailed records of the movements of Wardian cases—some of the few available showing how plants were moved in them.[6]

In 1841, when William Hooker first took over the directorship of Kew Gardens, Wardian cases began to be regular used. Hooker's administration spanned a period from the Gardens' first use of the Wardian to its

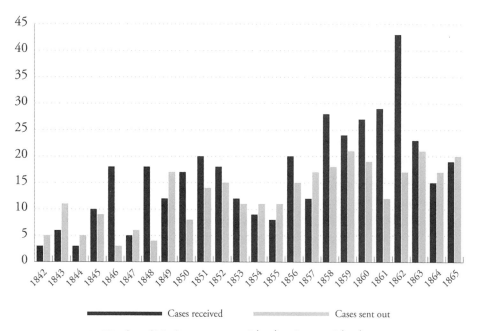

FIGURE 8.1 Number of Wardian cases, inward (399) and outward (307), Royal Botanic Gardens, Kew, 1842–65.

acceptance as a universal technology for moving plants. Hooker, as mentioned earlier, was close friends with Nathaniel Ward and first learned of the case when he sent George Gardner to Brazil in 1838 (see chapter 3). Focusing on the movements of Wardian cases throws new light on the well-known stories of transplanting economically valuable plants such as banana, tea, and cinchona. To move key economic crops, botanists first had to build a network of suppliers. And to do that, they had to learn about the diversity and the geographical distribution of plants.

Between 1842 and 1865 Kew Gardens sent out 307 Wardian cases.[7] On average they sent more than twelve cases per year, about one a month. The decade from 1855 to 1865 saw an even higher average, closer to twenty per year. Naturally, Wardian cases were not the only things sent by Kew. The most common consignments were seed packets, but they also included tin boxes, closed boxes, cuttings in boxes, and even glass bottles filled with seeds.

The Wardian case accounted for approximately 10 percent of all the packages sent by Kew during this time. This is a surprisingly high percentage, considering that most of the plants were sent within the United Kingdom to amateurs, such as Ward, and, importantly, nearby nurseries, including those of George Loddiges, Harry Veitch, Hugh Low, James Backhouse, and William Bull. They also sent shipments to nurseries on the Continent, such as Linden in Brussels and Haage & Schmidt in Erfurt, Germany, and to those further afield, such as Herbst & Rossiter in Rio de Janeiro. In 1856 Kew even sent boxes of flax, rhododendron, and camellia plants to Brackenridge's nursery in Baltimore (see chapter 4). Commercial gardens often served as distributors of new and rare plants, and many of these British nurseries turned a healthy profit.

The Wardian case was the primary method of moving consignments of live plants beyond Europe. Kew shipped cases all around the globe, but mostly wherever Britain had imperial interests: they sent a regular supply of plants to botanical gardens throughout the colonies. They also sent cases with a number of individuals: bishops such as John Colenso, headed to South Africa; explorers such as David Livingstone, going to central Africa; retired explorers such as Charles Sturt, returning to Adelaide, Australia; leading colonial officials such as Henry Barkly, posted to Melbourne; Rutherford Alcock traveling to Yeddo, Japan; and even amateurs and colonists, such as C. K. Williams in Bahia, Brazil, or William McLeay in Sydney.

If we take a look inside the cases, we also find a variety of plants. Kew used a number different cases, some were larger anywhere up to four feet long and three feet high and wide, but they could often be smaller, like the one that still exists in the Economic Botany Collection at Kew, which measures 70 cm W × 50 cm H × 83 cm D. They contained between thirty and sixty plants, both ornamental and useful, of at least twenty or thirty different varieties. Rarely was a case planted with just one species. Only around the 1860s, at the time of the cinchona expeditions, did it become common for cases to be sent with a single species inside. Interestingly, in 1862 Kew received a large number of Wardian cases from Japan that had been handmade by local carpenters. These cases were then reused, filled with cinchona and sent to the colonial gardens in the

West Indies. The Wardian case, it seems, was in short supply at Kew, and the Japanese carpenters had crafted some very well-built cases.

Even more than cinchona, ornamentals were by far the most popular plants sent out: roses, fuchsias, jasmine, camellias, gardenias, azaleas, chrysanthemums, and rhododendrons. The huge number of such plants sent suggests that a case full of ornamentals was a necessary cargo to take to colonies of settlers. But these gifts were given with the expectation that the cases would be returned filled with valuable local plants. Well before the cinchona project, whole cases containing nothing but roses were sent to settlers: in November 1853 William March in Jamaica received a box with thirty-seven varieties of roses, and in 1857 a case containing twenty-seven varieties of roses, and nothing else, was sent to Panama.

Kew sent far more Wardian cases containing British garden plants than of economic plants. Sometimes there was a mix: a box containing ornamentals might include a useful fruit tree or a banana plant or tea plant. Many plants from different parts of the globe were also sent out. The caladium, from Brazil, was sent to botanical gardens in Ceylon, Trinidad, and Mauritius. It was also common to send *Ixora* (from India) and others, such as specimens of the enticing *Aristolochia* and *Anthurium*. Kew processed what came in and then shipped it to foreign lands as an interesting exotic. The diversity of plants that were sent around the world was extraordinary; however, many of them became invasive in their new environments. Notable examples include the yellow flag iris (*Iris pseudacorus*), a major problem in Canada, the United States, and New Zealand, and the yellow jasmine (*Jasminum humile*), now strangling native plants in New Zealand.

Over the same period, 1842–65, Kew received 399 cases, ninety-two more than were sent out—an average of seventeen cases per year arriving from distant locations and a surplus of almost 25 percent. Surprisingly, over these two decades the number of cases that arrived fluctuated greatly: in 1862 forty-three cases arrived at Kew, yet in 1864 only fifteen were received. The number of cases that landed in London was reliant on many factors: weather, diplomacy, shipping, and the skills of the gardener packing the box.

The losses were high. The Wardian case failed in one out of four inward packages, and a quarter of the cases arrived with no living plants at all. On some routes this mortality was greater than on others. Nearly all cases that traveled overland from India or Ceylon to Kew arrived with all plants dead. It took the Melbourne Botanic Gardens two years (1858–60) to land a case of living plants, and even then the plants in seven of the nine cases sent to Kew arrived dead. The success of landing New Zealand plants was similar to that experienced from Melbourne, whose outward-bound specimens often arrived in "wretched condition."[8]

Why the failures? There were many reasons: the glass broke, or the plants were overwatered before packing, or they were too young. Many failures were the result simply of the uncertainties of shipping. In 1859 two cases arrived from Java; one was in very good condition, but in the other all the plants were dead, "having by accident been dropped into the Thames."[9] Despite claims to the contrary, the Wardian case was not a sure means of sending plants. As Joseph Hooker once wrote to his New Zealand correspondent, "We have hitherto been most unfortunate and I now call Ward's cases 'Ward's coffins!'" To be sure, at the time Hooker was desperate for New Zealand plants to arrive in London while he was completing his flora of the island, but all of the plants kept dying on route.[10] One other reason for the high failure rate was simply that most of the people sending the cases were amateurs.

The losses can also be attributed to the network that William Hooker operated. From very early in his directorship he used donations and distant collectors, a practice he had started in Glasgow, as opposed to outfitting plant explorers. And if he did send a plant explorer, he operated the expedition as a syndicate, often sharing the load with nurseries to help defray his expenses. These collectors in the distant colonies were often amateurs who collected in their spare time and had little experience in packing the cases correctly; instead of payment, Hooker supplied them with gifts and educational items. As Joseph Hooker wrote to his father in 1843, "You are quite right in not recommending the Crown to send out a Kew collector. It is more expensive than is worth while. . . . There are so many cheaper and as good ways of getting plants."[11]

The returning cases largely contained native plants from the regions

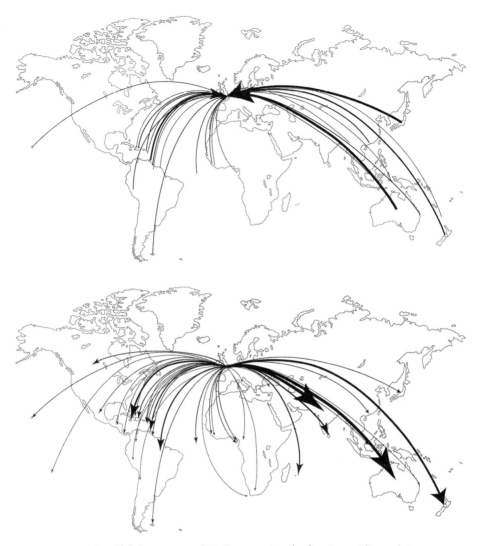

FIGURE 8.2 Global movement of Wardian cases into (*top*) and out of (*bottom*) Kew
Gardens, 1842–65.

of their dispatch. There were usually upward of twenty-five species in
each case, and there could be as many as sixty plants. In some cases there
was only one variety, such as the two Wardian cases that arrived from
Calcutta conveying the talipot palm (*Corypha umbraculifera*), one of the
world's largest species of palm (fig 8.2).

The plants came from all over the globe, but mainly from the colonies. The largest number of cases arrived from India, Australia, and Japan, respectively. But of the plants sent out from Kew, Australia received the most. The top three receiving countries were Australia, India, and Jamaica (closely followed by New Zealand). Most surprising is Japan: the first cases did not arrive until 1860, but over the next five years a large haul of plants was sent back to Kew. This coincided with the opening of Japan to foreign trade. As noted above, some of the cases made in Japan were reused, filled with cinchona plants, and sent to other locations.[12] Two of these cases were sent to Dominica and were later returned full of native Caribbean plants.

Even by just focusing on this short period at Kew, we see that the trade in plants was extensive and complex. It covered the globe, and Wardian cases flowed both into and out of Kew. They contained many varieties of both useful and ornamental plants. In the late 1850s, to keep track of the cases being sent, Kew decided to label their cases. The journeys of Wardian case no. 10 that began this chapter help to illustrate the worldwide use of Wardian cases.

One in seven cases coming into Kew was a returned case. This meant that constantly sending out cases was an important aspect of maintaining the supply. The example of the Japanese-made Wardian cases used to distribute cinchona in the 1860s also highlights the fact that the Wardian case was in short supply and at the same time vitally important in maintaining this logistical network. Many years later, when Joseph Hooker took over from his father as the director of Kew, there were still many issues with having enough Wardian cases. Hooker wrote a memorandum and sent it out to the places they supplied.

> The Wardian cases made specially for the Royal Gardens, Kew, and branded "Home Government Property" are not intended to be used except for the interchange of plants between Kew and the Colonies and India. As only thoroughly sound cases are sent out from Kew capable of lasting several years the Director will be much obliged if correspondents will be good enough to reserve these cases for Kew exchanges and return them the same year if possible, with such plants as are available at the time.[13]

During the directorship of Joseph Hooker (1865–85) the average number of cases received grew steadily to approximately twenty a year, a figure that remained steady up to the close of the nineteenth century.

Of the few records that still exist, the Goods Inwards/Outwards Books show both the distance and diversity of plants moved in Wardian cases. This trade served the interests of both imperial scientists and settlers. Scientists had a steady supply of new and rare plants from distant locations that they could use in their botanical gardens and to compile their botanical research in order better to understand the global distribution of plants. And settlers received ornamental plants from Britain.[14] In the global plant trade, the rose was as much a key plant as were bananas, tea, and cinchona.

Looking closely at the records of just one period at Kew reminds us that it was at heart a garden, one that sent out beautiful plants for others to start their own gardens in the colonies. To be sure, we are seeing here just two decades of Kew's plant movements; but these indicate that the bulk of the work done was in distributing and processing plants. Sometimes, however, there was a single species of economic value that was in high demand by imperialists, and at these times Kew used its experience to facilitate plant exchanges.

Seventy-Four Thousand Seeds, Thirty-Eight Cases

Natural rubber is an unstable product. It is sticky, easily loses its shape, and is brittle when cold. It starts out as a white sap — called latex or caoutchouc — that is tapped from specific types of trees. The product was a curiosity until 1839, when the process of vulcanization, effected by the application of sulfur and heat, allowed rubber to remain elastic. This unlocking of the material's potential set off a rubber boom. In the 1870s rubber was tapped from many different wild trees in South America. Once it was known that other varieties of trees yield rubber, many were tapped in Africa and Asia. Tapping rubber became a worldwide practice. The raw substance was collected by local inhabitants while they were performing other functions of daily life, traded in the local economy, and sold to traders. The most sought-after of the trees was the Brazilian

Pará rubber (*Hevea brasiliensis*), source of the purest and most elastic latex. Brazil was the leading producer. It was not long before plans were developed to take Pará rubber plants from their native home in Brazil to plantations in Asia. The Wardian case played an important role in the transfer.[15]

The British led the way. The effort was not without precedent. The apparent success of the cinchona transfer inspired botanists and imperialists to try to replicate it. Clements Markham, who worked at the India Office and was a key actor in the cinchona transfer, soon put his mind to work on rubber. Over a decade after he had reported to the Society of Arts on his cinchona exploits, he returned to lecture on rubber: "The time has come when plantations must be formed of caoutchouc yielding trees, in order to prevent their eventual destruction, and to provide for a permanent supply."[16] Much more than a concern for conservation, he was more troubled that the British had no control over such a precious resource. Markham worked closely with Joseph Hooker to put collectors in the Brazilian Amazon to source seeds and plants.

Numerous collectors were sent, but they had limited success. *Hevea* has oily seeds that are difficult to transport. Only a few seeds brought back by collectors could be propagated in Kew's greenhouses. In the early 1870s Kew recorded that barely 1 percent of seeds survived the journey from South America. In 1873 two thousand seeds made it from Brazil to Kew, but only twelve plants germinated. A Wardian case with six of these plants was sent to Calcutta Botanic Garden, but none survived planting in either Bengal or Sikkim. With this failure, Ceylon (now Sri Lanka) became the preferred site of rubber plantations. But they still needed viable Pará rubber plants.[17]

Henry Alexander Wickham was a self-promoting British colonist living in the Amazon Basin. Before leaving London, Wickham convinced many in his extended family that they would make their fortunes on the plantations he planned to set up on the Amazon. After many failures and the deaths of his mother and sister, he turned to collecting plants. In an attempt to raise funds, he offered to collect plants for Joseph Hooker. At first Hooker ignored Wickham's letters. But with the failures of other collectors, Hooker proposed, in a collaborative project with Markham at the India Office, that Wickham collect *Hevea* seeds, offering him £10 per thousand seeds. Early in their correspondence Wickham even proposed growing rubber seedlings on his Amazon plantation and sending them directly to Ceylon in Wardian cases. This idea was rejected as too costly. But the Wardian case was to become an important part of the project.

Wickham set about collecting in early 1876. On various canoe trips up the Amazon, and with help of many local inhabitants, he acquired 74,000 seeds. In May he left by the steamer *Amazonas* bound for Liverpool. From there the seeds traveled to Kew, arriving in mid-June. The assistant director at Kew, William Thiselton-Dyer, worked with the gardener Richard Lynch on propagating the seeds. It was a large operation, covering nearly three hundred square feet. The seeds were sown into flats and left in the greenhouse. They waited.[18]

"What we did was done in the most ordinary and routine way," remembered Thiselton-Dyer many years later. Three days after sowing the seeds, Thiselton-Dyer entered the propagating house and saw to his surprise that some of them had already sprouted. In total, however, only 2,700 of the seeds sprouted—a mere 3.6 percent. Kew had thirty-eight "special cases made" for shipping the plants to Ceylon. This was nearly as many cases as Kew usually prepared in two years. The cases were planted with 1,919 *Hevea* seedlings. Less than two months after arriving at Kew, on 9 August 1876, the cases were placed on board a P&O steamer under the care of a gardener and sent to Ceylon.[19]

When they arrived in Colombo on 13 September, it was found that nearly all of the precious seedlings had survived. Surprisingly, in such a well-organized operation, freight charges were not paid by the India Office. The cases sat on the docks in Colombo while Markham organized payment. Fortunately, the plants survived the delay. They were planted out at special hinterland locations at the Royal Botanical Gardens, Peradeniya, in Ceylon. Two of the thirty-eight cases were marked for trials in Singapore. The operation took only five months, from the time the seeds left Brazil until the young plants arrived in Ceylon. Such a fast and successful transplant could only have been effected with the use of the Wardian case.

Less well-known in the rubber-transfer story is that at more or less the same time as Wickham's operation, Robert Cross also went to the Amazon to collect rubber seedlings and bring them back to Kew. Strangely, Cross left Liverpool a few days after Wickham's seeds arrived. Nevertheless, once he got to Brazil he quickly commenced collecting. He first consulted a local collector, Don Henrique, who informed him of the best area in which to collect, and Cross set out with a local boy as a guide.[20] He gave instructions for the local people in Belem, a town along the Amazon, to make for him four Wardian cases in the meantime. In a little over a month Cross had amassed over a thousand good plants, which he sowed into the cases. He returned by the steamer *Paraense*, stopping along the way to collect other varieties of rubber-bearing trees. He arrived back in England on 22 November 1876. The plants were still alive but in poor condition, and Kew decided to keep only four hundred of them. The remaining six hundred were given to the commercial nurseryman William Bull. The following year Bull offered *Hevea*, "an important plant from the commercial value of its product," among his new, rare, and desirable plants to be sent to all parts of the globe in his specially designed cases.[21]

Over the coming years more shipments of both live plants and seeds were sent to Ceylon and Singapore. Most of these were from Wickham's seeds, but some came from Cross's supply. There is some speculation that while Wickham's plants provided the main genetic stock of rubber plants, the infusion of Cross's plants allowed for more successful cultivation. By 1882 the trees in Ceylon were providing seeds, and Kew was no longer involved in the transfer of plants. Despite the successful growth of the plants in Ceylon, there was little enthusiasm among planters for rubber.

A few years later, in an ironic reverse transfer, the Ceylon gardens sent two thousand seeds back to Kew because, as the director of the gardens said, "no one here wants it." The experimental crop of one thousand trees in Singapore was also producing seeds. In the 1880s the trees were tapped for the first time, with excellent results. However, plantation rubber took hold slowly. In fact, in the 1890s it was not rubber itself that was turning a profit for Ceylon planters, but the sale of seeds and seedlings. Plants were sold to a wide variety of customers, including plantations in the Dutch East Indies; German colonies in Cameroon, Tanzania, and Samoa; and Portugal's colony in Mozambique.[22]

By 1896 the price dropped out of coffee, and many planters in the Malay peninsula took up rubber. Within five years twelve thousand acres had been planted, and by one estimate 1.5 million trees were growing, "presumably the whole being the progeny of the trees originally introduced by the Government of India." It prompted Henry Ridley (fig. 8.3), the director of the Singapore Botanic Gardens, to write to Kew: "Now that the cultivation is well started it may before long be expected to prove one of the most important products in the Colony."[23]

By the early twentieth century, when the rubber plantations in Ceylon were established, they were even supplying rubber to planters in Brazil. The plantations in Brazil suffered terribly from leaf blight, and genetic material from the new plantations in Asia offered possibilities, particularly to planters from the Americas, for a revived Brazilian rubber industry. The Brazilian rubber industry was struck on two fronts: first, by a severe attack of disease, and second, by the burgeoning Asian rubber-producing industry. It never recovered.[24]

In Ceylon and the Malay Peninsula two businesses cropped up. The first was tapping rubber on plantations, and the second was selling seeds and seedlings (fig. 8.4). Commercial nurseries supplied rubber seedlings in Wardian cases to colonists and agriculturalists throughout Asia. Although seeds could be purchased from the same nurseries, the case was often preferred among planters looking to establish crops quickly; this was particularly true of German planters. Some consignments of Pará rubber plants were for as many as 75,000 seedlings.

While Pará rubber was the most sought after, many varieties of rubber-bearing trees were valuable to Europeans and to global markets, such as gutta percha, which became one of the most important plant-based products of the late nineteenth century, used not only to coat undersea telegraph cables but also for toys, interior decorations, boots, water pipes, and dentistry (where it is still used today). The Wardian case was an important method for sending rubber plants to botanical gardens to have new varieties identified and tested for commercial possibilities.

Rubber is but one of many examples of what an important and suc-

FIGURE 8.3 Henry Ridley ("Rubber Ridley") standing next to a rubber tree with a local worker. Ridley was a Kewite, the director of the Singapore Botanic Gardens (1888–1911), and a leading promoter of the rubber industry in Southeast Asia.
© The Board of Trustees of the Royal Botanic Gardens, Kew.

FIGURE 8.4 A child worker in Ceylon stands with Wardian cases of rubber plants ready
for shipment to South America, ca. 1913. From H. F. Macmillan, *A Handbook of Tropical
Gardening and Planting*, 2nd ed. (Colombo: H. W. Cave, 1914), 638.

cessful technology the Wardian case was for colonists. It was not the only
way to move plants—seeds were the preferred method—but sometimes
live plants had to be moved. With rubber we see how both methods were
used in a chain of connection from Brazil via Kew to Ceylon and Singapore. Following the successful shipping of seeds to Kew, Henry Wickham's services were no longer required. As a parting gift for his services,
Joseph Hooker gave Wickham two Wardian cases filled with Liberian
coffee. Wickham took these with him to Queensland, in northern Australia, where he again tried and failed to set up a successful plantation.

The Case of Kew

William Thiselton-Dyer, who oversaw the rubber project at Kew, was a
man of empire. Among his many achievements, he became the president

of the International Rubber Planters' Association. In 1885 he took over from Joseph Hooker as director of Kew. He is remembered for promoting Kew as an indispensable center of colonial botany. The Colonial Office, the Foreign Office, and the India Office all relied on Kew's expertise. Kew was consulted in the search for new or improved vegetable crops and usually advised on the most suitable crops for plantations. Kew also trained many of the gardeners and botanists who took up colonial outposts—affectionately known as "Kewites."

In the late nineteenth and early twentieth centuries, the imperial project supported by Kew continued. Plantations took over more land. In the final decade of the nineteenth century, 1891–1901, Kew moved 198 Wardian cases. The cases went all over the world: Accra, Boston, Calcutta, Ceylon, Fiji, Hong Kong, Java, Lagos, Melbourne, Natal, Penang, Sierra Leone, Singapore, St. Vincent, Sydney, and Trinidad. Between 1890 and 1910 Kew sent an average of about twenty cases per year.[25]

The Wardian case was successfully used for nearly a century at Kew. Unlike those of many other botanical gardens, Kew's records provide specific insight into how the global plant trade operated. In the mid-nineteenth century many ornamental varieties were sent out and many native plants sent back. But by the late nineteenth and early twentieth centuries the importance of plantation crops, particularly for imperial projects, meant that the flow of plants was often about moving economic crops. The success of the rubber transplant, in which the Wardian case played a key role, was an example that many other nations attempted to replicate. Today rubber is ever-present in our daily lives. After its successful transplant to Asia in the 1870s, the former colonies in Southeast Asia became the world's leading rubber producers, a status they hold to this day.

PLATE 1 Merina men carrying plants in Wardian cases from the Nanisana botanical station to a countryside plantation, Madagascar, ca. 1899. Photo by Emile Prudhomme, © CIRAD.

PLATE 2 Edward Hopley, *The Primrose from England*, ca. 1855. Oil on canvas. In the 1850s, three thousand people lined the streets to see the first primrose, sent in a Wardian case, after its arrival in Melbourne, Australia. Hopley painted the scene after hearing the story of the floral traveler. Collection of the Bendigo Art Gallery, gift of Mr. & Mrs. Leonard V. Lansell, 1964.

PLATE 3 The Wardian case is a rare object in museums today. Shown here are four of the very few remaining plant boxes from around the world. The two on the left predate Ward's invention. *Clockwise from top left:* traveling plant case used by the Frenchman André Thouin to move coffee, ca. 1781, © Muséum national d'Histoire naturelle, Paris; the Wardian case used by Kew, ca. 1890, Economic Botany Collection, © The Board of Trustees of the Royal Botanic Gardens, Kew; Wardian case of the Waroona Historical Society Museum, Western Australia (this case was only very recently discovered, and for a long time the former owner used it as a dog kennel); and the plant and seed box used by Allan Cunningham, ca. 1795, collection of The Royal Botanic Garden and Domain Trust, Daniel Solander Library, Sydney.

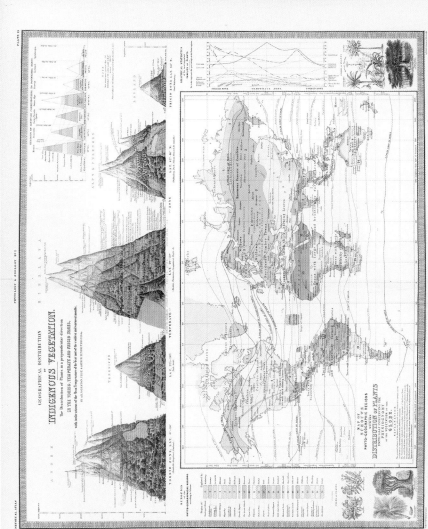

PLATE 4 "Geographical Distribution of Indigenous Vegetation," 1856. Many plants were moved so that scientists could map the biogeography of plants around the globe. From Alexander Keith Johnston, *The Physical Atlas of Natural Phenomena* (Edinburgh: William Blackwood & Sons, 1856). Courtesy of the David Rumsey Map Collection.

PLATE 5 The French d'Urville expedition navigating Adélie Land, Antarctica, 1840. Drawing by Léon Jean-Baptiste Sabatier. Collection of the National Library of Australia.

PLATE 6 Indoor-style Wardian case. This beautiful case was made by Andrew Brown in Glasgow between 1860 and 1880. Planting display by Aaron Angell for the Gallery of Modern Art, Glasgow, 2017. Collection of the Glasgow Museums. Photo by Max Slaven.

PLATE 7 Interior view of a typical European glass factory. Increased glass production was an important factor in decreasing the cost of manufacturing Wardian cases. This one was in Bavaria, Germany, ca. 1889. Painting by Rudolf Wimmer, collection Deutsches Museum, Munich, Archives, BN52869.

PLATE 8 Seymour Joseph Guy, *The Contest for the Bouquet: The Family of Robert Gordon in Their New York Dining-Room*, 1866. Oil on canvas. Notice the Wardian case on the windowsill, to the left. Collection of the Metropolitan Museum of Art, New York.

PLATE 9 Gardeners at the Jardin d'Essai Colonial prepare Wardian cases for transport, 1909. D.R., © CIRAD.

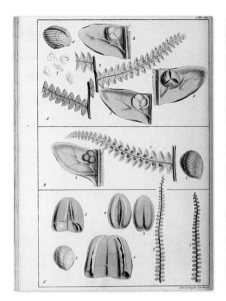

PLATE 10 Four of the many beautiful plants moved in Wardian cases. *Clockwise from top left*: the first fern to arrive in Britain in a Wardian case, *Gleichenia microphylla*, illustrated by Franz Bauer, from W. J. Hooker, *Genera Filicum* (London: Bohn, 1842); the plant John Gibson went looking for in India in 1836, *Amherstia nobilis*, from *Curtis's Botanical Magazine* (1849); Japanese umbrella pine, *Sciadopitys verticillata*, taken from Japan to the United States by George Rogers Hall, from Louis van Houtte, *Flore des serres et des jardins de l'Europe* 14 (1861); and *Daphne fortuni*, "the harbinger of spring," one of the many ornamental plants Robert Fortune moved in Wardian cases, from Louis van Houtte, *Flore des serres et des jardins de l'Europe* 3 (1847).

PLATE 11 Wardian cases preparing to leave the Buitenzorg botanical gardens, Dutch East Indies, Indonesia. Collection Nationaal Museum van Wereldculturen, Image TM-1001076o.

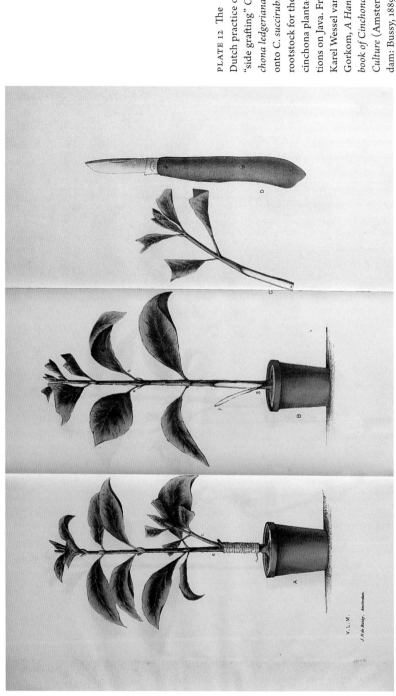

PLATE 12 The Dutch practice of "side grafting" *Cinchona ledgeriana* onto *C. succirubra* rootstock for their cinchona plantations on Java. From Karel Wessel van Gorkom, *A Handbook of Cinchona Culture* (Amsterdam: Bussy, 1889).

PLATE 13 Gates of the Adelaide Botanic Garden, Australia, 1877. The director, Richard Schomburg, an avid user of the Wardian, installed cases along the entry path to welcome visitors. Photo by Samuel White Sweet. Collection of the State Library of South Australia.

PLATE 14 Gardeners at the Burnley School of Horticulture, 1934. Far more than just indoor gardeners, women played important roles in many aspects of the garden trade. Collection of the Burnley Gardens, University of Melbourne Archives.

PLATE 15 Marianne North, *A New Pitcher Plant from the Limestone Mountains of Sarawak, Borneo*, 1876. After seeing North's illustration of the pitcher plant, *Nepenthes northiana*, the Veitch nursery firm in Chelsea sent collectors to Borneo to collect the plant. © The Board of Trustees of the Royal Botanic Gardens, Kew.

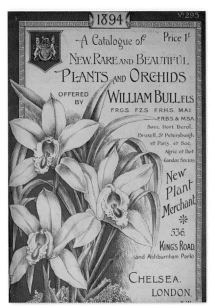

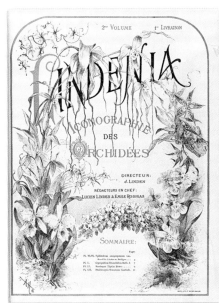

PLATE 16 Four catalog covers on which nursery firms advertised their newest and most beautiful plants. *Clockwise from top left*: A Catalogue of New, Rare and Beautiful Plants and Orchids Offered by William Bull (1894); *Catalogue of Seeds, &c: James Veitch and Sons, Royal Exotic Nursery, Chelsea* (1896); *Descriptive Catalogue of the Yokohama Nursery Co.* (1918–19); and Jean Linden and Émile Rodigas, *Lindenia: Iconographie des orchidées* (Brussels, 1886).

FIG. 1.—ORANGE COVERED WITH SOOTY MOLD (From Morrill and Back.)

FIG. 2.—LEAF OF ORANGE COATED WITH SOOTY MOLD (From Morrill and Back.)

PLATE 17   Orange and leaf after infestation of whiteflies. The sooty mold develops on the secretions of the whitefly. Solutions to the problem in Florida were found after "natural enemies" were transported from Lahore, India, to Florida, United States, in Wardian cases. From Russell Woglum, *A Report of a Trip to India and the Orient in Search of the Natural Enemies of the Citrus White Fly* (Washington, DC: Government Printer, 1913).

PLATE 18   The cactoblastis moth, *Cactoblastis cactorum*, feeding on prickly pear. The moths were moved to Australia in Wardian cases as a biological control for the prickly pear, which had become invasive. From Alan P. Dodd, *The Biological Campaign against Prickly-Pear* (Brisbane: Commonwealth Prickly Pear Board, 1940). Collection of the National Library of Australia.

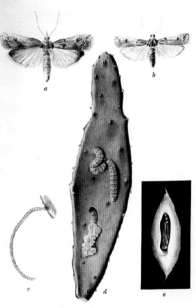

Cactoblastis cactorum.

(*a*) Female moth   (*b*) Male moth
(*c*) Eggstick   (*d*) Larvæ
(*e*) Cocoon showing pupa

[*By courtesy of the*
Queensland Department of Agriculture and Stock.

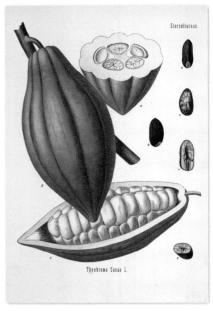

PLATE 19 Four useful plants that were moved in the Wardian case. *Clockwise from top left*: Cavendish banana, *Musa cavendishii*, from *Paxton's Magazine of Botany* 3 (1837); tea, *Camellia sinensis*; cacao, *Theobroma cacao*; and rubber, *Hevea brasiliensis*. The final three images are from *Koehler's Medizinal-Pflanzen* 2, no. 3 (1883–98).

9

Case of Colonialism

In the late nineteenth century, when many political leaders were pursuing a civilizing mission and firms were looking to increase profits, agriculture was a central part of the colonial project. Moving plants to plantations required Wardian cases. By the late nineteenth century, knowledge of plants and their distribution extended to most parts of the globe. At the same time, the profits of firms and plantations operating in colonies were tied to a thorough scientific understanding of which plants could be cultivated in certain locations. Botany played an important role in the profitability of colonies. As important as understanding where plants might be transported was the possibility that they could be moved at all.[1]

During this period technological developments in Europe were an important ingredient in the spread of imperialism. A range of technologies—including steamships, railways, telegraphs, and repeating rifles—aided Europeans in taking their imperial mission to many parts of Africa and Asia. Other factors, such as faster shipping routes, also played an important role. The Suez Canal was opened in 1869, and by 1910 freight rates fell by half. The cases that sailed on these ships moved around the globe at greater speed than ever. Looking at the Wardian case, we see an older technology, over seventy years old by the end of the nineteenth century, that found new uses with the expansion of empires. The Wardian case and other older technologies were not just a ready means by which to move plants but also a link to earlier periods of informal imperialism. Where explorers and travelers and distant collectors from an earlier period sent an odd box or two, now botanical gardens set in motion hundreds of Wardian cases for the express purpose of establishing colonial and commercial plantations.[2]

"New Imperialism" is the label historians give to this intense period of formal imperialism between the 1880s and the beginning of World War I. British, Dutch, French, Portuguese, and Spanish colonial interests had evolved over many centuries, but in the late nineteenth century renewed interest in state-sponsored expansion emerged with Belgium, Germany, Japan, and the United States. All of these countries sought to control the natural resources of other countries and continents, and setting up agricultural plantations was a key part of all these imperial powers' projects. By setting up plantations, colonists could turn unused land into productive land and along the way appropriate native peoples into working that land. European botanical gardens became centers of colonial botany, distributing plants and information to the colonial botanical gardens and research stations and also, more often than not, to the firms that operated there.

This chapter continues the story of colonial botany. It tells the story of how the Wardian case was packed and sent by German and French scientists in the late nineteenth century. During this period the case was an important technology for moving plants, in particular to set up colonial plantations and also to search for new plants. The Wardian case was not only a prime mover of live plants for colonists, but also, as we will see, an important symbol of the work being done in both gardens and the colonies. The field of colonial botany in the age of New Imperialism is vast and global, so in this chapter the focus is on two German and French colonial institutions that were very successful in moving plants. As we will see, in the first decade of the twentieth century the French moved more cases than Kew.

The increase in the movement of Wardian cases in the late nineteenth and early twentieth centuries had serious impacts on indigenous people

FIGURE 9.1 Colonial cacao plantation in Cameroon, 1923. Photograph Pacalet, © CIRAD.

and indentured workers as land and labor were appropriated into a global plantation economy. We are still living with the consequences of these plantations. For example, cacao, from which the primary ingredient in chocolate is sourced, was sent to the colonies in Wardian cases (fig. 9.1). Today the Ivory Coast, Ghana, and Cameroon—former French, British, and German colonies, respectively—produce more than 70 percent of the world's cocoa. In building their colonial botanical enterprise, many nations looked to Kew as inspiration.

Transplanting Kew

Plants were important to imperial aspirations. Agriculture served two purposes. First, it allowed imperial powers to carve out large portions of foreign territories and in doing so bring native peoples onto the plantations as forced labor. This was about control of both land and people and was promoted by imperial countries as part of their civilizing mission. Second, in the late nineteenth century colonizers, such as Britain, France, and Germany, argued that plantations were an important developmental tool and that this justified the colonization of distant lands and native territories. Kew's work was well-known to other imperial powers in Europe. For other nations, replicating the work of Kew was an important aspect of their colonial projects.

Efforts toward a German colonial botany started in the 1880s, with colonial societies seeing plantations' potential. As in many other aspects of building their empire, the Germans looked to Britain as an example. Botanical information had been flowing between Kew and Berlin for many years. As early as 1879 the director of the Berlin gardens went to Kew and was given "all possible facilities" by Joseph Hooker. But in 1888 two German visitors arrived who were not botanists. They had been sent by the Deutsche Kolonialgesellschaft (German Colonial Society).[3] The Kolonialgesellschaft was founded in 1887 and within six months had made arrangements for two young members to investigate the work being done at Kew. The Kolonialgesellschaft had large influence throughout Germany and at the highest levels of government, and it pushed for greater expansion of the German empire, an enterprise to which a thorough understanding of botany was key. By sending two members to Kew, the they were preparing to, "create an institution in Germany, which should bear similar relations to the German possessions abroad, as the Kew Gardens bear to the British colonies."[4]

In the ensuing years a correspondence sprang up between William Thiselton-Dyer, the director of Kew Gardens, and Georg Schweinfurth, a well-known German explorer in Africa and founding member of the Deutsche Kolonialgesellschaft. Schweinfurth understood the important role Kew played in transplanting plants to their colonies and believed that if Germany wanted to do the same it needed a similar botanical institution. The Kolonialgesellschaft sought detailed and specific information from Kew, not just on the gardens, but on Kew's "interior arrangements." The Germans were quite ambitious in wanting to create another Kew, not just in its grandeur of the gardens but in its colonial reach and administration.[5] After consulting with those above him in the Office of Works, who were responsible for Kew, and the Foreign Office, Thiselton-Dyer decided instead to send the Kolonialgesellschaft copies of some official publications, such as Kew's *Bulletin of Miscellaneous Information*. In a personal letter to Thiselton-Dyer, Thomas Villiers Lister, from the Foreign Office, wrote skeptically about the German botanical enterprise: "The Germans should understand that for an institution like Kew to flourish it should, like a big tree, have a very small beginning. Transplanted full grown it is not likely to succeed."[6] Nonetheless, the Germans pushed ahead with their plans for a central botanical establishment. And they were not alone.

In 1898 the French sent a botanist to Kew to speak with Thiselton-Dyer. They wanted to learn more about Kew's work in the colonies. Following this meeting, in September they prepared a report on the work at Kew detailing how it could be replicated. The following month the Minister of Colonies organized a committee to investigate such a network of test gardens. On 28 January 1899 the French president issued a decree setting aside a portion of the Bois de Vincennes, the largest public park in Paris, in the eastern suburb of Nogent-sur-Marne, for the Jardin d'Essai Colonial.[7]

Both the French and the Germans now had colonial botanical gardens to service their imperial aspirations.

Berlin: A Collection as Complete as Possible

The Botanic Garden Berlin was one of the oldest and most well-known in Europe. It had been laid out in the seventeenth century on the Potsdamerstrasse, close to the center of Berlin. One of the earliest gardens was the hops garden, which supplied a local brewery. Interestingly, in the 1830s William Brackenridge worked at the Botanic Garden Berlin before he migrated to the United States and joined the Wilkes expedition; the size of the collection, especially the succulents and ferns, impressed him, but the small grounds neither provided extensive views or room for expansion. By the 1890s the rapidly expanding city surrounded the gardens and there was little room to expand, so they were moved to a much larger site at Dahlem, on the city's fringe. The move came at a time when botany was assuming an important role in the expanding German empire.[8]

In the late nineteenth century the German Reich emerged, with ambitions to spread over the globe. These were mostly pushed by commercial interests looking for state support in their overseas dealings. Starting in 1884, around the time of the Berlin Conference, Germany acquired territories in Africa, Asia, and the Pacific. By 1891 the Botanische Zentralstelle für die deutschen Kolonien (Botanical Research Center for the German Colonies) had commenced operations from what is now known as the Botanic Garden and Botanical Museum Berlin.

German colonial ambitions offer a fascinating glimpse of the intense period of New Imperialism. Although appearing late in the game, German colonialism was quite typical of its time. The historian Ulrike Lindner has commented: "On many points German colonizers acted similarly to their imperial neighbours in the colonies and were meshed into a tight network of transnational and transcolonial interaction."[9] This colonizing mission was almost always about expanding economic interests and controlling the resources and land of distant territories. For the Germans, it was also about acquiring the prestige accorded to other imperial powers. The plantations in Cameroon, Namibia, Papua New Guinea, Samoa, Togo, and Tanzania were all vital to the spread and justification of German colonialism.

Following Germany's acquisition of colonial lands came making those lands productive. Clearing forests and setting up plantations was one of a number of ways to effect this step. For the colonists, plantations allowed unused land to be made productive and often native peoples were forced to work the land.[10]

Adolf Engler was appointed director of the Berlin garden in 1889 and oversaw the colonial botanical research project. Engler was an energetic scientist not personally taken with the colonial project but ambitious enough to understand that holding strong connections to the Foreign Office would be of great benefit to his garden. In the pages of the Berlin garden's own publication, *Notizblatt des königlichen botanischen Gartens und Museums zu Berlin* (Notes of the Royal Botanical Garden and Botanical Museum, Berlin), which was modeled on Kew's bulletin, Engler wrote: "With a collection of tropical plants as complete as possible in the garden and a rich collection of plant products in the museum, the institution shall offer everyone who leaves for the colonies the opportunity to become familiar with these things."[11] At the new garden in Dahlem, Engler arranged for all the economic and interesting plants from the colonies to be placed in one location so that visitors could imagine the productive work Germans were doing in their colonies. In its first season, in 1903, over five thousand people visited the garden to see coffee, peanut, tobacco, tea, rubber, and sugar plants.

The Botanic Garden Berlin served as an important distributor of plants. By the turn of the century there were botanical stations in the German colonies of Victoria, Cameroon; Amani, Tanzania; Missahoe and Sokodé, Togo; and Simpson Harbour, New Guinea. But it was not always like this. An examination of the garden's work before and during the colonial period shows that it was transformed by its new role in the colonial project. While in 1880 only five hundred live plants arrived at the gardens, by 1894 this had increased nearly fourfold.[12]

In 1889 the first Wardian case was sent with the express purpose of meeting German imperial interests. The magazine *Kolonie und Heimat* (Colony and Homeland) reported that "the first task of the [*Botanische*] *Zentralstelle* is to supply the colonies with seeds and live plants. The sending of these occurs in the so-called Wardian cases."[13] The magazine reported that "hundreds of boxes with many thousands of young plants of all species have been delivered to East Africa, Cameroon, Togo and even to the South Sea." The article concluded by telling readers that many of the plants sent to these distant colonies "literally originated in Berlin" (fig. 9.2).

The first Wardian cases went to the Victoria botanical garden in Cameroon (now Limbe Botanic Garden) in 1889 and formed its foundational collection. It became an official colonial botanical garden in late 1891. The next consignment of Wardian cases arrived from Berlin in the summer of 1892. The garden also received plants from other locations. Wardian cases containing cacao plants came directly to Cameroon from Trinidad, and within a short time 332 cacao saplings were thriving in the gardens. More

FIGURE 9.2 Wardian cases in Berlin ready for use, 1910. From *Kolonie und Heimat,*
17 July 1910.

plants flowed from Berlin in April 1893, August 1894, and October 1896.
Often Wardian cases were sent out under the care of a trusted person
who could oversee the plants on the voyage and in doing this the Victoria garden reported very few losses.[14]

The Wardian case was an important mover of plants for the German
colonial project. By the turn of the century it was the most important
method of moving plants to the colonies. As Engler in Berlin reported
in 1901, "Living plants in Wardian cases [were] sent to the experimental
gardens in Victoria and Buea in Cameroon, Lome Little Popo in Togo,
and the Sigi plantation company in Usambara."[15] In 1902 Berlin sent fifteen cases containing over a thousand plants to Tanzania, and another

eleven cases to the other colonies. Similar practices continued over the next decade. German plant hunters also used the case effectively. Following in the footsteps of Henry Wickham (see chapter 9), Ernst Ule, on his expedition to Brazil, procured rubber plants in the key rubber area of Manaus and not only collected seeds but also sowed those seeds into Wardian cases and sent them to Berlin. Plants were moved to all the research stations and botanical gardens in the colonies, and even to the German consul in Nicaragua. The Botanic Garden Berlin also sent plants to German companies operating plantations in the colonies, including the German-English East Africa Company, the German Rubber Society, the German East African Plantation Society, the German West African Plantation Society, and the New Guinea Company. The plants on the move—a total of at least seventeen thousand—included coffee, oil palms, cacao, rubber, vanilla, gutta percha, bananas, pineapples, mangoes, papayas, sisal, and rubber, as well as many ornamentals. By 1910 a total of 240 Wardian cases had been sent by the Germans to their colonies. Half of these were sent to Cameroon, 64 to East Africa, 56 to Togo, and 20 to the South Sea colonies.[16]

Thousands of plants were put on the move to satisfy the desires of people in the colonies. Most were economically important plants that could be grown and processed for profit. Images from the magazine *Kolonie und Heimat* illustrate how both in Europe and in the colonies the movement of plants required the efforts of many individuals. While there were grand plans by the German Reich and by the Deutsche Kolonialgesellschaft, with their commercial interests, the Wardian case shows that the success of the plantations also relied on the simple work with plants.

Samoa became a German colony in 1900, but German firms, such as the Deutsche Handels- und Plantagen-Gesellschaft (German Trading and Plantation Company), had been operating on the island since at least 1879. By 1900 the nearly five thousand acres of the trading and plantation company on the island of Upolu were proclaimed the "largest single complex of tropical agriculture in any of the German colonies." Germany took control of Samoa after much pressure from the plantation

FIGURE 9.3 Workers unpacking Wardian cases sent from Berlin to Samoa, 1911. From *Kolonie und Heimat*, 21 May 1911.

companies. One of the main issues that they faced was the availability of labor. Following colonization, German diplomatic channels opened discussions with China to send workers to Samoa.[17]

Figure 9.3 reveals a number of important details. These laborers are unloading what looks like a shipment of either cacao or rubber seedlings; both were important crops on the German-controlled island of Upolu. A closer look shows Chinese men unpacking plants. Over two thousand Chinese indentured laborers were brought to work alongside other indentured Melanesian workers in Samoa. Conditions in Samoa were harsh, wages low, and workdays over eleven hours, and sometimes they only had two Sundays off per month. Severe penalties were dealt for disobedience, but Chinese workers nonetheless often protested their conditions.

The Deutsche Samoa Gesellschaft was known as a brutal employer that issued horrific punishments. A plantation manager named Richard Deekin, enraged by a mass petition written by Chinese workers, beat one of them, a man named Ah Tsong, so severely that four of his friends had to carry him to town for medical treatment. Deekin was charged with the offense. In Germany, however, Deekin was also a well-known booster of plantation life, having written a number of popular books, such as *Manuia Samoa! Samoan Travel Sketches and Observations* (1901), promoting Samoa to German immigrants. Conditions were so terrible under the German plantation owners that in 1908 the Chinese government stopped laborers from going to Samoa, although the practice recommenced the following year.[18]

Returning to the image of the Chinese and Melanesian laborers unpacking the Wardian cases in Samoa, there is also another detail that deserves attention. The Wardian cases sent out by the Botanic Garden Berlin were modeled on the designs of the Chelsea nurseryman William Bull. One of these cases still survives in the Botanic Garden and Botanical Museum Berlin (see fig. 1.2). Here we have a major botanical institution distributing thousands of plants to the colonies, and the technology being used was copied from a nurseryman in Chelsea. It is a strange mix of colonial science and civil society.

The Botanic Garden Berlin moved thousands of plants and helped German colonists set up plantations in many parts of the globe. But it must not be forgotten that this came at a cost. In Africa, conflict between native people and colonists was common, sometimes resulting in such horrific massacres as the Herero and Namaqua genocide. Many companies operating in Africa were given a warning because of their cruel treatment of workers and native people.

While the Germans sought to control both people and land, some kinds of control were much harder to achieve. After importing coffee trees from Ceylon, German East African plantation owners noticed coffee leaf rust on the plants (*Hemileia vastatrix*). Interestingly, coffee rust is known to have originated in nearby Ethiopia, the natural home of coffee, and was first transferred to Ceylon by British plant movers. It then returned to Africa with German colonists in 1894. In 1913 the disease spread across Africa from Kenya to Congo, wiping out large and valuable plantations.[19]

On 15 August 1914, just after the outbreak of the World War I and only weeks before the fall of Samoa to New Zealand, the *Samoanische Zeitung* reported on "the colonies" of Germany. The article made an important point about the spread of plantations in the three decades of the German colonial empire. In German East Africa there were 45,000 hectares of rubber plantations, 24,000 hectares of sisal, nearly 13,000 hectares of cotton, 8,000 hectares of coconuts, and nearly 5,000 of coffee. In Cameroon the "European plantings" amounted to 13,000 hectares of cacao, over 7,000 of rubber, 5,000 of palm oil, and 2,000 of bananas. The list went on to cover the colonies in Togo and New Guinea and concluded with Samoa, where there were nearly 5000 hectares of coconut palms, nearly 4,000 of cacao, and over 1,000 of rubber. Germany's colonies were lost with the fall of the empire in World War I and divided among other nations. In a brief three decades the Germans had gained control of large portions of foreign land. Plants were important in this process, and this was aided by the ability to move them successfully in Wardian cases.[20]

Paris: The Colonial Garden

In 1897, from the colonial outpost in Tunisia, in North Africa, the respected French naturalist and African explorer Jean Dybowski wrote a short pamphlet titled *Les jardins d'essai coloniaux* (The Colonial Test Gardens). The pamphlet outlined Dybowski's vision for a network of colonial test stations that were linked to a central test garden in Paris. This was not another pleasure garden, but a practical one. His statement set the tone for French colonial botany for decades to come. The French were well behind the Germans in organizing their colonial botany. Although they already had agricultural departments set up in their possessions, they had no organized center. As Dybowski claimed, and many imperialists in France agreed, an organized set of botanical test stations, controlled and led from a central Paris location, was important for colonial expansion.[21]

Dybowski wrote his pamphlet while director-general of agriculture in Tunisia, a post that gave him a different perspective from scientists based in Europe. He was the archetypical colonial promoter. In 1891 he explored the French Congo with the aim of expanding trade and outposts in the region. When his fellow explorer was murdered in a small village, Dybowski apparently took it upon himself to avenge the death by inflict-

FIGURE 9.4 Students at the Jardin d'Essai Colonial prepare Wardian cases for transport, 1934. D.R., © CIRAD.

ing the same pain on members of the village.[22] The role of the test garden in Paris that he proposed was to cultivate profitable economic species and send them out to colonists. Unlike the scientists who focused on taxonomy and nomenclature at such institutions as the Muséum national d'Histoire naturelle, Dybowski wanted a colonial garden that would focus on profitable plants. Such a station would save the newcomer to the colony precious time, thus "hastening the moment of the first harvests."[23] The Wardian case was an important mover of plants in these moments of haste.

The French organized quickly. They sent an envoy to see the work at Kew, and by January 1899 they had a decree issued by the president to set up a new colonial garden in the eastern Paris suburb of Nogent-sur-Marne. Thus was born the Jardin d'Essai Colonial (fig. 9.4), now known as the Jardin d'Agronomie Tropicale (Garden of Tropical Agriculture). Greenhouses and buildings went up at Nogent-sur-Marne, and the work of supplying the colonies started immediately. One of the earliest images that exists of the colonial test garden shows the newly built greenhouses and also shows one of the gardeners waving and standing over a Wardian case packed and ready to send (see fig. 4.1).[24]

The French had considerable experience with Wardian cases, which, ever since the first ones arrived in Paris, had been used not only by the natural history museum but also by a number of French nurseries. The Muséum national d'Histoire naturelle, the leading French museum, issued detailed instructions to travelers, and by 1865 the Wardian case was the primary means of moving live plants. At this time the museum adopted a slightly smaller case. By the time of the colonial test garden, the Wardian case was very similar to the designs promoted by William Bull and used by the Germans. At the same time French nursery firms, such as Vilmorin-Andrieux, were promoting "special packaging" to overseas colonists, telling them of the advantages of using "glass boxes (Ward boxes)" for shipping seedlings. In 1898, when the Muséum national d'Histoire naturelle became involved in setting up the colonial test garden, it also began sending transfers of gutta percha in Wardian cases to the French Congo.[25]

By the late nineteenth century, with the push for agricultural development in the colonies and the establishment of the colonial test garden, there was even greater movement of live plants. For the French, this operated in two ways. First, they had to classify the local plants and map the countryside to understand how best to exploit local products. Second, once they knew the biogeography of a place, they also knew which imported economic plants could be sent from the central test garden. Madagascar is an example.

The plants of Madagascar were of great interest to many botanists, including the British.[26] The French gained control of Madagascar after two bloody wars with the native population. In 1897 the French colony was declared with the fall of the Merina kingdom. The French immediately set up a garden in the capital city of Antananarivo and a botanical station outside of the city near Nanisana.

In 1896 the botanist Emile Prudhomme became director of agriculture on the island. His first order of business was to exploit its native plants. One product that very much interested the French was local rubber: on Madagascar they could extract rubber from at least thirteen different local species. In the final three years of the nineteenth century, rubber from local trees was their biggest export, making it the island's first export crop. When it arrived at the Antwerp exchange, one of the varieties was labeled "East Coast Niggers," a name that shockingly describes the colonial mind: rubber was accepted into the global market, but local people were cruelly derided. There were other plants of interest in Madagascar, such as the fibrous bark used by the Antemoro tribe for making paper, or the *Raphia* palm, used at the time for grafting. There were also many native medicines. For the French, finding useful plants was vital to their colonial project. For the Merina, not only were their lands now under control of the French, but they were coerced into cultivating and moving the plants of their country to best serve the colonial rulers.[27]

Few other images in the archives are as evocative as the one of four Merina men carrying Wardian cases (see plate 1). Here we see colonist and colonized in the business of plants. The movement of plants could never have occurred without the labor and knowledge of local people. Only a short time after the fall of the Merina empire, these men were filling and moving cases for the white men.

From the very first years of rule in Madagascar, the "search for improvements" in the ways of growing plants was at the forefront of French minds. As Joseph Gallieni, the military leader who overthrew Merina rule, noted in his 1899 report on colonization, the introduction of plants was a key task of the new colony. Furthermore, he wrote, the French people working at setting up plantations in the colony needed to be supplied with the seeds and plants they required.[28]

Once the rootstock of valuable plants was set up at botanical test stations, it was distributed to the surrounding plantations. Space was dedicated on the Nanisana grounds for raising plants sent from Paris. There were also test gardens set up at Ivoloina, in the east of the country, and Nahampoana, in the south. Plants were then distributed from these government gardens by the Department of Agriculture. The main varieties included cacao, pepper, Pará rubber, nutmeg, and several varieties of the vanilla orchid. Sometimes specific orders were placed by plantation owners. These were then filled in Paris, raised at the test stations, and delivered. Only upon delivery did plantation owners pay the costs. In 1902 the agriculture department in Madagascar distributed an estimated 150,000 young live plants throughout the colony.[29]

The French had vast colonial possessions. At their height they covered eleven million square kilometers and included over one hundred million people, with colonies in North Africa, West Africa, Equatorial Africa, the Indian Ocean, and Indochina. All of these locations had botanical test stations that sent and received plants, just as in Madagascar. Rubber plantations in Indochina became vitally important in the early twentieth century. From the beginning, cacao in the Ivory Coast was an important plantation crop, as were peanuts in Senegal. Whether it was live plants, seeds, or new strains of economic crops, all of the test stations in the colonies were linked to the central garden in Paris.[30]

In Paris, just as in Berlin, the Jardin Colonial was a showcase for would-be colonists. The test garden also became a teaching institution where young agricultural students were trained specifically for work in the colonies; Dybowski became one of the teachers. After his stint in Madagascar, Emile Prudhomme became director. Significantly, the garden was to become the site of the 1907 Exposition Coloniale.

The Exposition Coloniale transformed the Jardin Colonial into a microcosm of empire. The Exposition was as much about people as it was about plants. Laid out in a series of pavilions, all specially transported from the colonies, each contained both people and buildings. There was the *village indochinois*, with people from Vietnam and Laos brought to Paris and even a rice paddy growing for the display. Other pavilions included exhibitions from Congo, Caledonia, Guiana, Madagascar, and Tunisia, all with specially imported buildings to give the feel for the colonies. What started out as a garden for the cultivation of plants had now become a site for displaying the exotic. During the Exposition the greenhouses at Nogent-sur-Marne remained in use, and between 14 May and 6 October 1907 the Jardin Colonial welcomed more than two million visitors to the gardens. This was not just a particularly French exercise: the Dutch, Germans, and Belgians all held similar exhibitions showing off their colonial possessions. The most surprising aspect of the exhibition at Nogent-sur-Marne was that the buildings brought over from the distant colonies remained at the site for the use of scientists in the future. Many are still there today, all of them in ruins. The historian of the French empire Robert Aldrich writes that the site is arguably the most "poignant colonial lieu de mémoire in France, their empty buildings the shells of empire."[31]

A few years after the Exposition the journalist Louis Vernet visited the site and described it with all the colonial zeal of the times in the journal La dépêche coloniale: "While you cross the Vincennes woods, whose flora and sights are solely those of the French forest, you will indeed be on French soil. You will feel at home. But when you turn the corner, as soon as you see the Jardin Colonial, immediately the impression will strike you that you are entering a different part of the universe, that you are reaching new lands. The landscape is no longer the same. You see giant bamboos, tall pines, all sorts of greenery that are not from here, all sorts of buildings that are not from here either. This is Africa. This is Asia. This is exoticism. You are in the colonies."[32]

He also toured the test gardens. There were greenhouses and propagating houses, papaya trees with fruit, seedlings of cacao ready to send, even a field of sweet-smelling geraniums planted outdoors. Most notably, he saw a large horse-drawn cart being loaded with Wardian cases starting their journey to the colonies.

Not only was it important to see the cases in Paris; Vernet included with his article images of the test station at Ivoloina in Madagascar (fig. 9.5). They showed a young Malagasy boy in a field of new cacao trees and a white planter next to a much taller cacao tree. They also showed the economic plants in the test gardens at Ivoloina, with the caption: "Most of the economic plants assembled with the Ivoloina were sent in portable greenhouses [Wardian cases] through the Jardin Colonial."[33] The image is revealing: the Malagasy with their children on the left and the white colonist on the right, also with her children, the two groups segregated by a line of trees.

The Wardian case became a point of particular pride for colonial agriculturalists. In the French Pavilion at the Brussels Exposition Universelle

FIGURE 9.5 The Botanical Station in Ivoloina, Tamatave. From *La dépêche coloniale*, 30 June 1909.

et Internationale of 1910, the French were able to show off the economic-plant products they had sent to their colonies: "improved" varieties of cacao, coffee, rubber, and fruit. The exhibit told visitors that the colonial test garden in Paris had shipped more than forty thousand specimens of live economic plants in the "so called Wardian cases." In the decade since its creation, 1899–1909, the Jardin d'Essai Colonial shipped at least four hundred Wardian cases to the colonies, almost twice as many as the Germans and certainly more than the British over the same period. With the help of the Wardian case, they had transformed their colonies.[34]

Expanding Empires

The Berlin and Paris examples were not the only colonial gardens, there were others in Hamburg and Lyon. The Dutch were also significant movers of plants who had centers of colonial botany in Amsterdam and Leiden but relied heavily on their Buitenzorg gardens in Indonesia and on commercial nursery firms. The Russians in St. Petersburg moved large numbers of plants as well, and the Belgians moved many ficus and coffee plants from their center at the Royal Laeken greenhouses. The United States Department of Agriculture, as we will see in the next chapter, was also heavily involved in moving plants. What we see in the examples of the French and Germans is that they really did adapt the Kew model quite effectively for the purposes of their colonial ambitions. One important aspect in their success was in using the Wardian case to move live plants.

Plants have always been important for empire building. But in the late nineteenth century valuable crops such as rubber, cacao, coffee, coconuts, peanuts, vanilla, bananas, and palm oil became increasingly important. Many of these plants were transferred in Wardian cases. We are still living with the consequences today: cacao is the Ivory Coast's biggest export, Senegal's main export crop is peanuts, Tanzania's coffee industry is a key part of its agriculture sector, and rubber is primarily produced in Asian states that were once British and French colonies.

The Wardian case was vital in establishing much of the rootstock for many plantations, but there were, of course, many follow-on impacts even after its use had ceased. The French offer a case in point. The next phase of agricultural research, covering the period between the two world wars, was a time of specialization. As the historian Christophe Bonneuil has written: "Each colony was to specialize in the large-scale production of a few priority agricultural products, to supply mainland France."[35] For the experimental stations discussed in this chapter, even the German stations that were taken over by French and British interests, specialization often meant that each location focused solely on one particular crop. This later period set up many of the agricultural monocultures we have today.

Many of the plantations set up in the late nineteenth and early twentieth centuries were very important in setting the tone of agriculture for many nations in the postcolonial period. The Wardian case was used for over half a century before the new imperialists came along. But when they did, they packed these old cases full of new plants to take to their new colonies. In this intense phase of colonial activity, the Wardian case underwent a reinvention to become a key technology in the expansion of empires. Successfully used by Kew Gardens and quickly picked up by other institutes of colonial botany, it was a ready-made method that facilitated colonial expansion. But as we will see in the next two chapters, ecologies are never so simple. Moving plants had many unforeseen consequences.

10

....

Burning Questions

To move live plants was to move ecosystems. The inside of a Wardian case was a micro-ecosystem of plants, soil, water, and captured sunlight, and with transpiration it functioned a little like our larger, outside world. It is no coincidence that during the peak of its use, the spread of plant diseases and invasive species also grew dramatically. The many examples, as we will see in the next two chapters, are a strange mix of weeds, worms, and beetles. After almost a century, the years of uncontrolled movement of species was coming to an end.

Environmental management emerged in the early twentieth century as a leading paradigm of the day. Very different from how we think of it today, much of this early work set out to tame nature according to human whim, to make it more tractable.[1] But as this approach was carried to agricultural economies, in cohabitation with imperialism, a whole series of unintended and unanticipated consequences was unleashed. Consider the monoculture plantations that the British, French, and Germans set up with their colonial projects. Once these plantations were established, it was not long before plant diseases and insects arrived to ravage the crops. While the Wardian case had been an important technology for distributing plants, it had also brought many smaller organisms along for the ride. A number of these diseases and pests were challenging the very agricultural economies that the cases had spread so widely. We had become very good at moving environments—too good.

As early as 1881 European nations gathered together to curb the flow of unwanted species into Europe.[2] The main concern was to stop the spread of *Phylloxera vastatrix* (today known as *Daktulosphaira vitifoliae*), an insect pest that ravaged grapevines. Arriving on California vines, the pest was first discovered in France but soon infested vineyards all over Europe and later Australia, New Zealand, South Africa, and Peru. The Convention signed at Berne, Switzerland, in April 1889 was the first of many international agreements to stop the spread of invasive pests, diseases, and fungi. In the beginning these agreements targeted individual problems. Interestingly, until the 1920s these agreements did not affect the flow of Wardian cases or the soil inside them. Ultimately these international agreements culminated in the wide-reaching International Plant Protection Convention, adopted in Rome in 1951.[3]

With faster and more regular shipping, plants were moved in ever greater numbers. Not only was there a rapid increase in the movement of plants around the globe, but, paradoxically, restricting those movements then became a fundamental part of environmental control. One of the biggest movers of plants in the early twentieth century was the United States Department of Agriculture (fig. 10.1). No institution or user displayed the paradoxical environmental management of the time more than the USDA. One of its bureaus, the Office of Foreign Plant Introduction, was responsible for bringing in tens of thousands of plants. But another, the Bureau of Entomology, saw the devastation caused by foreign introductions, particularly insects, and led the campaign for plant quarantine.

By the 1920s the movement of Wardian cases had become increasingly controlled and restricted. Although the United States came late to controlling the flow of species, the laws and regulations it eventually enacted were the most widespread and extensive among all nations. Government permits were required, and nearly all shipments had to be impounded temporarily at one of the designated quarantine stations. There were many complex tensions between introducing new plants and quarantine, and it was in this context that the Wardian case finally met its

FIGURE 10.1 P. H. Dorsett inspecting Wardian cases aboard the *Utowana*, 1932. On this trip David Fairchild supplied plants to the United States Department of Agriculture. Also visible in the background is the flag of the United States. Collection of the Archives of the Fairchild Tropical Botanic Garden.

demise. This chapter focuses on the United States, its use of the Wardian case in the early twentieth century, and the competing tensions within the USDA.

"Great Oaks from Little Acorns Grow!"

In the United States the work of foreign plant introduction cannot be overstated: today there are actually only a few crop plants of commercial value that are native to the United States.[4] In 1898 the Section of Seed and Plant Introduction was established by the USDA with the mission of finding, testing, and distributing useful foreign plants. Like many governmental divisions, it went through many name changes, and in the 1920s it was relabeled the Office of Foreign Plant Introduction (a name used throughout the rest of the chapter). By this time the United States was importing between three thousand and four thousand plants annually.[5] In the early twentieth century the Office became an important player in the global movement of plants. Unlike in the previous chapter, where we saw imperial powers processing plants through their main botanical institutions and sent them *out*, through the Office of Foreign Plant Introduction the United States brought plants *in*.

David Fairchild, a botanist and plant explorer, was the enigmatic figure who led the Office of Foreign Plant Introduction from 1904 to 1928.[6] He was acutely aware of the imperial importance of plants. In 1898, just as the Office's work was starting, Fairchild wrote: "The rapid development of any new country is due to the discovery of soil and climatic conditions suited to the growth of introduced food plants, and seldom to the development of an endemic species."[7] For Fairchild, the agricultural success of grapes and oranges in California was a prime example: "So thoroughly has this [success] been recognized by all colonizing nations that they have established botanical gardens in their new colonies, one important function of which is to secure and distribute exotic economic plants throughout the colony." In the United States, this work was set to increase with the organized efforts of Fairchild's division.

Many plants had been introduced into the United States before the Office was set up. Remember Robert Fortune was contracted by the Patent Office in 1859 to bring tea from China (chapter 5).[8] Following the establishment of the Department of Agriculture in 1862 by President Abraham Lincoln, there were many early efforts to introduce and test valuable plants. There was also the ever-popular congressional seed distribution service, which disseminated vegetable, field crop, and flower seeds.

As we have seen, one of the earliest formal attempts was made during the Wilkes expedition. This early introduction was not lost on Harley Bartlett, a former employee of the Office, who wrote during the expedition's centenary celebrations: "The present vast work of the Department of Agriculture's Plant Introduction Office is just as clearly to be interpreted as an outgrowth of what Brackenridge started at the little Patent Office greenhouse. Great oaks from little acorns grow!"[9] On the Wilkes expedition (chapter 4) the horticulturist William Brackenridge filled Wardian cases with exotic plants and brought them to Washington. It took over half a century to fully realize Brackenridge's vision for a "receptacle for proving all samples of fruits, flowers and succulents,"

FIGURE 10.2 F. G. Harcourt and assistants posing with a Wardian case at Dominica
Botanic Gardens, Roseau. The plants pictured are "American economic plants" taken
by David Fairchild to Dominica and were exchanged for interesting economic plants,
rare palms, and frangipani. Part of the Allison Armour expedition, 1932. Photo by P. H.
Dorsett. Collection of the Archives of the Fairchild Tropical Botanic Garden.

but when the American plant explorers finally did, they certainly fulfilled
Brackenridge's vision.

The Office of Plant Introduction was the first formal section to sys-
tematically introduce and test plants for introduction into the United
States (fig. 10.2). The Office's plant programs had an imperial focus,
and for the Americans there was an overriding interest in the botany
of the tropics, where many of their imperial projects were located—a
biogeography where the Wardian case was an important tool. American
and Russian botanists were leaders in the science of plant breeding. Getting access to wider genetic stock was an important part of creating new
strains of existing agricultural products.[10]

The Office's work was extensive. Their ongoing *Inventory of Seeds and Plants Imported* was an epic list of plants that detailed the yearly introductions carried into the United States. By 1928, after three decades, the inventory listed nearly eighty thousand plant introductions.[11] These accessions were global, as the Department admitted: "Nearly every country in the temperate regions of the earth has yielded its store of worth-while things."[12] To be sure, the overwhelming number of imports were of seeds, but the Wardian case was a regular and important mover of many plants.

Among the many plant introductions Fairchild oversaw during his administration of the Office were the flowering cherry, pistachios, nectarines, bamboo, avocados, East Indian mangoes, and horseradish. Fairchild, a great promoter, was well connected: his father-in-law was the inventor Alexander Graham Bell. He regularly promoted the work of the Office in beautifully illustrated articles for such popular publications as *National Geographic, World's Work,* and *Youth's Companion.* Fairchild's vision was romantic, exotic, and cosmopolitan. Indeed, many of his promotions appear more in step with the nineteenth century than with the institutionalized environmental management of the early twentieth century. But Fairchild was always quick to point out the practical benefit of his introductions. An article in *Popular Science Monthly* based on Fairchild and his work could have easily been found inside the pages of *Kolonie und Heimat* or *La dépêche coloniale,* with its images of Wardian cases in the tropics being packed and readied for transport.[13]

One of Fairchild's overriding interests was to increase the variety and diversity of foods and fruits available to Americans. One of his favorite fruits was the mangosteen, "the queen of tropical fruits," as he described it, which required Wardian cases.[14] Describing the "remarkable" collection of Chinese vegetables imported by the plant explorer Frank Meyer, Fairchild wrote: "The Chinese restaurants which are scattered all over the country are creating a taste among Americans for these new vegetables, and the next step in their introduction will be their culture on a small scale to supply the growing demand of these restaurants."[15] In 1932 one media outlet reported that whenever you ate an avocado, bamboo sprout, lemon, or bread made from durum wheat, you could thank the

Office of Foreign Plant Introduction.[16] The introductions that Fairchild oversaw continue to impact all Americans.

Just as the Office's work was beginning, American scientists were starting to pay much closer attention to the tropics. There were major imperial projects in the tropics with the Panama Canal and expanded imperial interests in places such as the Philippines. Many of these projects and places required a thorough knowledge of these humid and, as many believed, threatening environments. Environmental knowledge and management were key elements in successfully "conquering" the tropics, and a thorough understanding of tropical botany was critical for these efforts.[17] For the people involved in plant introduction, it also meant that there was a ready supply of government officials deployed to these locations who might turn out to be useful collectors. Moving plants from the tropics was difficult for the Americans. As Fairchild explained: "The seeds of many tropical plants are very short lived. If dried they die and if kept moist they germinate in a few days, so that the only way to send them is as seedlings in Wardian cases."[18] The Wardian case was thus important for transporting the tropics to the United States.

Wardian cases were used for the introduction of numerous plants, both economic and ornamental. The list of its uses is diverse and includes a case of essential-oil grasses from Ceylon, cases of grafted lychees from Canton deposited in Hawaii, and cases from China of different species of *Garcinias* (related to the mangosteen). Between 1909 and 1913 the United States received many plants from the Buitenzorg botanical gardens in Indonesia (now the Bogor Botanical Gardens). From the Dutch colony there were varieties of mango and magnolia, *Atalantia monophylla,* and the delicate fruit-bearing *Feroniella lucida.* In one consignment from the new colony of the Philippines, the director of the forestry department sent two hundred plants of four different varieties of mangroves packed in cases. From Ceylon they received a case of mangosteens "carefully brought back" to the United States on the steamer *Aloha* under the supervision of the collector Mrs. Arthur Curtis James, who had secured it from gardens in Peradeniya, Sri Lanka. Such acts of transport by faithful collectors were impressive: "The Office has made it a point never to miss an opportunity to secure new stock, whether in the form of a shipment of seeds by parcel post, or a wardian case of young plants which some traveler returning from the East has generously volunteered to bring home." The case was used by many correspondents to send plants to Washington. While some of these were potential agricultural products, many of them were also exotic horticultural varieties.[19]

The Office's plant explorer Wilson Popenoe offers a good example of the Wardian case in use. While exploring in India and Sri Lanka, Popenoe sent mangoes and pears in Wardian cases, as they "can only be had in the form of plants."[20] Popenoe worked closely with A. C. Hartless, the director of the Royal Botanical Gardens, Peradeniya, to find new and interesting varieties, taking "copious notes" from Hartless's "fruit files." The results of gathering both plants and information greatly contributed to Popenoe's *Manual of Tropical and Subtropical Fruits* (1920), a pioneering American work on tropical agriculture, which described various tropical fruit crops worth growing in the United States. The mango is described in detail, including the emerging plantations in Florida. Popenoe neglected to mention that many of these were imported in Wardian cases.

The wider circulation of both plants and knowledge was important. Popenoe gained his start as a plant explorer searching for avocados in central America; indeed, his name is remembered today in the common name of the Popenoe avocado. In India, Popenoe not only took notes from Hartless's material, but he also shared information on his work on avocados. Hartless wrote to Washington: "If the Wardian case of avocado would come . . . it would be utilized in sending the plants he [Popenoe] selected." Hartless asked that a Wardian case of avocado plants be sent to him; once he received it, it would be returned to Washington with fruit selected by Popenoe.

Many of the Wardian cases arrived before 1912. After that things began to change in the USDA. By 1916, when Fairchild was arranging a collection of lychees from China, he had to defend the use of Wardian cases. The lychee has short-lived seeds and were difficult to transport, so Fairchild wanted to ship them in Wardian cases. Frank Meyer was the plant explorer employed to collect the plants in China, and as Fairchild was quick to point out, Meyer was not only "unusually expert in shipping

FIGURE 10.3 Boxes of plant cuttings and seeds leaving Beijing, China, collected by the explorer Frank Meyer of the Office of Plant Introduction, 1913. © President and Fellows of Harvard College, Arnold Arboretum Archives.

and packing plants" but also had "keen powers of observation when it comes to diseases of trees" (fig. 10.3).[21] With the lychee expedition, and many others that followed, the Office had to "make sure only healthy trees are imported."

The cost of importing so many plants was adding up. And while Fairchild and others in the Office thought there were benefits of importing and exploring, many were becoming increasingly cautious.[22]

Building a Wall, or an Open Door?

On the afternoon of 19 January 1917 the entomologist Charles Marlatt took to the stage in the conference room on the top floor of Washington, DC's grand New Willard Hotel, wedged serendipitously between the White House and the Old Patent Office Building. He was addressing the International Forestry Conference on the topic of "Stopping Importation of Tree and Plant Pests." The movement of plants was coming to a dramatic close. Looking out the window of the famous hotel, one would have seen the lawns of the National Mall in the distance, which nearly eighty years earlier had been covered by greenhouses filled with foreign plants ready to be distributed throughout the nation. The scientists gathered were members of the American Forestry Association, the United States' oldest conservation organization, and they were largely in agreement that most of the pests affecting crops and forests came from imported nursery stock.

Marlatt commenced his talk by listing the "vast horde" of disease-causing fungi, bacteria, and insects introduced into the New World: *Cryphonectria parasitica* (responsible for chestnut blight), *Cronartium ribicola* (white pine blister rust), *Xanthomonas axonopodis* (citrus canker), *Limantria dispar* (the European gypsy moth), *Anthomonus grandis* (the boll weevil), and *Icerya purchasi* Maskell (the cottony cushion scale). Although he did not specifically list the method of introduction, we now know that many of these were closely linked to the Wardian case. He told the audience, "The total loss occasioned by these introduced pests to our national forests and farm crops, etc., probably exceeds $500,000,000 annually."[23] In today's dollars that is about $9.8 billion—an extraordinary cost. As the head of the USDA's Bureau of Entomology and chairman of the Federal Horticultural Board, Marlatt was both respected and connected. That same year scientists in Marlatt's branch at the USDA would release *A Manual of Dangerous Insects* (1918), which listed over three thousand insects that could drastically affect American forests and agri culture. Some of them had already been introduced in Wardian cases.[24]

Many entomologists and plant pathologists were calling not just for stricter quarantine but for a complete ban on all nursery stock imported into the United States. Such measures would heavily impact the importations of plants by scientific institutions, botanical gardens, and even the Office of Plant Introduction. Quarantine officials already knew that many pests and diseases had already made their way into the country, often on nursery stock. The only way to eradicate such pests was incineration: burning the plants, the soil, and the Wardian cases. Marlatt concluded with a particular reference to plant explorers and importers of exotic material. While it was desirous to "accumulate from the ends of the earth interesting novelties and curiosities," the real priority must be "the conservation of the big commercial crops of America such as wheat, corn, cotton, potato, apple, peach, orange, etc."[25]

Marlatt was followed at the podium by a young botanist from the Office of Foreign Plant Introduction. P. H. Dorsett read the paper on behalf of his boss, since Fairchild probably knew he would be the cat among the pigeons.[26] The paper noted that many nurserymen and importers were concerned by the situation. But they were not the problem; rather, it was the big department stores, which imported large quantities of cheap bulbs and evergreens. Variety and diversity were fundamental for breeding new crops. Refuting the rhetoric of the entomologists, he described their policies as "building a wall of quarantine regulations."

Dorsett concluded with both practical and scientific observations. In a practical sense, the increasingly integrated global economy would not work with strict quarantines: "The whole trend of the world is toward greater intercourse, more frequent exchange of commodities, less isolation, and a greater mixture of the plants and plant products over the face of the globe."[27] On the specific science of plant breeding, which required imports of foreign germ plasm, he noted: "The restriction of the breeder and the nurseryman in the species which he would have at his disposal would tend to limit his activity and his interest and slow down the process of the production of new forms." In Fairchild's view, much of the enterprise of the Office of Foreign Plant Introduction worked on this premise: their search was not just for new plants, but for new varieties that would benefit the US economy. Indeed, Dorsett himself would go on to import varieties of soybean that would lay the foundation of the American soybean industry. Ultimately, however, the weight of Marlatt's evidence was hard to refute.

The movement of plants had so many unknown consequences that large restrictions needed to be put in force. We know this today as the "precautionary principle"—if the potential consequences of an action are too risky, then we need to stop that action. At the forestry conference in 1917, a resolution was passed that the American Forestry Association was in favor of an "absolute national quarantine" on imported plants and nursery stock that was to take effect as soon as "economically expedient." It was not just the foresters to whom Marlatt spread the message; he had talked to and gained support from many other groups.

Plants were first regulated in the United States with the introduction of the Plant Quarantine Act, passed by Congress in 1912. With this legislation, imported plants were inspected and if necessary quarantined; the main targets were nurseries, plant enthusiasts, and botanical gardens.[28] The outcome of this regulation was to establish the Federal Horticulture Board. Within a few short years the Board, with Marlatt at its head, had acquired substantial funding and influence.

In 1919 the most dramatic regulation was put into effect. Quarantine 37, as it was called, extended the regulations on imported plant material particularly by private interests, severely curtailed the ability of commercial nurseries to import ornamentals, and drastically reduced the flow of Wardian cases.[29] While the 1912 act required only that much of the imported material be inspected by officials, Quarantine 37 added a thick layer of bureaucracy to the process.[30] Most nursery imports of exotic plants required permits, and the process of obtaining them was a lengthy one; furthermore, if the United States already had enough plants that could be propagated, then the permits were often denied. Any plant material brought in had to go through the USDA and its quarantine stations. The new quarantine regulation set out whole categories of plants, mainly nursery stock, that were not allowed. Sweeping and controversial in its intent, Quarantine 37 set the foundation for the quarantine regulations in force in the United States today.[31] Many of these policies were picked up by other nations around the world, including Australia and New Zealand.

Following the 1919 legislation Marlatt published numerous articles in popular magazines. One of them, in *National Geographic*, presented readers with some compelling evidence. He wrote that before 1912, when there were no restrictions, "America became a dumping ground for the plant refuse of other countries."[32] On another page Marlatt showed plants with intact roots. "Moist earth," Marlatt wrote, "has been the source of a host of our worst plant enemies, such as Japanese beetle, the alfalfa weevil, and many others"—moist earth, like that in Wardian cases. With such large numbers of plants and seeds on the move, a strict and wide-reaching law was needed to put a stop to the imports.

For many years Fairchild and Marlatt were close friends. But the quarantine debate divided the plant explorer and the entomologist.[33] What is even more fascinating is that although the Wardian case was the means by which many of these pests arrived in the United States, it was again called upon to transport species that would control the pests. Indeed, when the entomologist Russell Woglum traveled to India in 1913 to find biological controls for the citrus whitefly, he developed new techniques using the Wardian case to transport insects. Woglum noted, "The writer's attention was called to this case by Mr. C. L. Marlatt, assistant chief of this bureau, and Mr. David Fairchild in charge of plant introductions."[34] While they were debating the pragmatics of quarantine, both Marlatt and Fairchild were promoting the Wardian case to import both plants *and* insects.

Commercial nurseries, horticulturalists, and plant explorers protested the strict quarantine laws. In one of the more insightful commentaries on the policy, Stephan Hamblin, director of the Botanic Gardens at Harvard, wrote in the *Atlantic Monthly*: "If such a policy had been enforced a hundred years ago, America would lack at the present time more than half the plants that make her gardens beautiful and more than half the fruits, grains, and other economic plants that make her horticulture profitable and advantageous."[35] This was the strange point that surrounded much of this controversy. Many of the very crops needing protection had at one point or another been moved to the United States from elsewhere. Protest came from many areas. One of the more organized bodies was the Committee on Horticultural Quarantine, led by J. Horace McFarland, which was trying to lessen the impact of the strict regulations. It represented interests ranging from the Harvard University's Arnold Arboretum and the Missouri Botanical Garden (fig. 10.4) to the Garden Club of America.[36]

The quarantine regulations dealt a hard blow to commercial nurseries on the other side of the Atlantic. For many in Britain and Europe, the decision came suddenly and surprisingly. As the *Gardeners' Chronicle* said, it will "very seriously" affect the commercial nursery trade, which supplied many Americans.[37] The *Gardeners' Chronicle* was scathing of

FIGURE 10.4 Wardian cases full of cycads from Rockhampton, Queensland, arrive at the Missouri Botanical Garden in St. Louis after a long journey via London and New York, 1913. Courtesy of the Missouri Botanical Garden.

the Federal Horticultural Board, arguing that its regulations were un-scientific and unfairly targeted horticulturalists. The magazine claimed that before the Americans target foreign nurseries, they should come to terms with their own contradictory policies within the USDA. If the Americans really were serious about stopping pests, argued the *Gardeners' Chronicle*, they should stop the "admirable work" of Fairchild and his Office of Plant Introduction.

The matter was also taken up by the Royal Horticultural Society in London. They thought they could effect change using their political in-fluence. Many of their members were engaged in the international com-mercial nursery trade and were greatly affected by the new laws. At their monthly meeting in January 1919, the Society spent much time discuss-ing the new law.[38] They sent letters to the US ambassador to Britain, the Secretary of State, the president of the Board of Trade, and David Fairchild, recommending that they rethink the harsh new law. The letter was signed by the executives of the Horticultural Society: the president Lord Grenfell and Harry Veitch. They also used British diplomatic channels, lobbying the Foreign Office and trying to get the British ambassador to present their concerns in Washington. By their March meeting there was a reply from the Americans. It was unequivocal: the USDA would maintain its strict position on quarantine and importation. At that point the Horticultural Society felt that no more could be done.

While many in the nursery industry believed that strict regulations might soften, it was slowly realized that the nineteenth century's free-flowing plant trade was gone forever. Although some of the more draconian measures were relaxed, ultimately the strict regulations led to further institutionalization. Charles Marlatt became the head of the Plant Quarantine and Control Administration, which was set up to further control the flow of plants into the United States.[39]

Burning

In the twilight of his career the pioneering USDA plant pathologist Beverley Galloway penned a short pamphlet titled *How to Collect, Label, and Pack Living Plant Material for Long-Distance Shipment* (1924). The intent of the pamphlet was similar to that expressed elsewhere a century earlier: to instruct government foreign travelers on how to bring plants of value back home. Galloway was wary of Marlatt's wide-ranging quarantine polices, but he still warned against the Wardian case. The pamphlet began with a short note on the difficulties of long sea voyages through the tropics. It soon, however, went on to warn of the "greatest danger" of introducing insect pests or plant diseases. These pests and diseases, Galloway warned, caused losses of millions of dollars annually. Throughout his career Galloway understood the importance of introducing new genetic material for the development of crops. Despite the large burden of regulation, Galloway encouraged plant introducers, because new crop plants were critical for maintaining genetic diversity in existing crops. In the pamphlet we see the tightrope between quarantine and introduction that many in the USDA walked. Although Galloway tried to be both

FIGURE 10.5 Graham Fairchild, David's son, uncovering the Wardian cases on the SS *President Adams* during the Allison Armour expedition, 1926. The tarp was placed over the cases after a storm in the Indian Ocean. Photo by David Fairchild. Collection of the Archives of the Fairchild Tropical Botanic Garden.

cautious and encouraging, the most serious warning was given regarding the Wardian case: "The Wardian cases are always sources of much danger." He went on: "The Wardian case . . . has probably been the means of scattering more dangerous insects, nematodes, and other pests over the earth than almost any other form of carrier; hence its use is not advised except under special instructions."[40] The days of the Wardian case were numbered. In the previous century the Wardian case had been promoted as the most important means of transplanting plants. Now it was a source of danger, to be used only when no other method was available.

It is then somewhat surprising that on David Fairchild's final journeys as a plant hunter, the Wardian case was his key technology for moving plants. Between 1924 and 1927, and again from 1931 to 1935, Fairchild traveled with the wealthy philanthropist Allison V. Armour.[41] Either Fairchild was ignorant of the instructions of his mentor and other colleagues in the USDA or he really saw the Wardian case as an indispensable tool of plant exploration. He and Armour brought back thousands of plants (fig. 10.5; see also fig. 10.1).[42]

Plant quarantine was a divisive issue in the early twentieth century. It was used as a precautionary measure to save crops and valuable plant resources. But it also curtailed the commercial activities of the international nursery trade as well as scientists who were searching not only for new and interesting plants but also for diverse breeding stock. This is most evident in the United States, but over the coming decades other countries also battled with the issue. By 1951, when the multilateral International Plant Protection Convention was adopted, most countries around the world had implemented plant quarantine regulations.

The Convention marked the end of the Wardian case's reign. No longer could a plant explorer set out in search of plants, have boxes made in a distant location, and use them to ship those plants to employers at home. Now the boxes had to go through a special, increasingly complex process. First, they were fumigated numerous times along their journey. Then, upon arrival on distant shores, they entered one of the many quarantine stations set up to receive and test newly arriving plants. Upon arrival, all soil was removed from the roots, and the plants were then placed in special cages and set aside to grow. Often a cutting was made and grafted to new stock already planted in American soil.

Finally, the discarded soil and the case were incinerated. As one former USDA employee described it, "The wooden cages or wardian cases formerly used, had to be burned as soon as they had served for a single lot of plants."[43] By the 1920s Wardian cases were being used only once and then burnt. By the 1930s a large-scale program of burning Wardian cases became common. It was the end of the case and its long and useful history.

11

········

Wardian Cages

To move plants in Wardian cases required soil. The average-sized Wardian case, like those used at Kew, contained more than two cubic feet of it. Just one cubic foot of soil, biodiversity scientists suggest, contains millions of different microbial species.[1] Biodiversity is both the key to ecological systems and the ecological lifeblood of human health. As the renowned biologist Edward O. Wilson eloquently remarked: "When you thrust a shovel into the soil . . . you are, godlike, cutting through an entire world. . . . Immediately close at hand, around and beneath our feet, lies the least explored part of the planet's surface. It is also the most vital place on Earth for human existence."[2] Inside this cubic foot of soil could be bacteria, algae, nematodes, insects, and mites. Some of the most important creatures in soil are earthworms: by moving, they mix the soil, and by eating and excreting, they process organic matter. They loosen the soil and increase its organic content and nutrient-holding capacity.[3] In soil we see the abundant diversity of life.

At the same time that rubber seeds were planted and propagated in the orchid house at Kew, a unique discovery was made: a worm. This was no ordinary worm. Long and thin, with five distinct dark-violet stripes down its back, it was most remarkable for the strange flat head protruding from its body. Today it is commonly known as the shovelhead worm or the hammerhead worm. In 1876 many interesting discoveries were made in the soil of Kew's greenhouses. Having spotted of this worm numerous times, gardeners working in the greenhouses took a preserved specimen to the deputy director, William Thiselton-Dyer, who had the worm sent to Henry Nottidge Moseley at Exeter College, Oxford. Moseley, an expert on planarians (worms), announced it as a new species. To memorialize the location of its discovery, he named it *Bipalium kewense*.[4]

Kew's worm was a strange specimen. Like other flatworms, the shovelhead does not have a circulatory or respiratory system. The mouth also functions as the anus and is positioned in the center of the body. Not only its appearance but also its provenance was strange. Moseley recognized that the worm was foreign to British soil; in 1878 he noted, "Unfortunately it is quite uncertain from what region it may have come, since the house in which it was found contains plants from various parts of the world."[5] Originally naturalists thought it hailed from Australia, but later discoveries showed it had a native range extending from Vietnam to Cambodia.[6] Within a few decades it became naturalized outdoors and in some locations had detrimental effects on native earthworms.

Flatworms are generalist carnivores, meaning they will feed on anything from earthworms to snails, and they often go after prey much larger than themselves. Flatworms pose a major threat to native earthworms. Removing native earthworms leads to major disruptions in local soil ecology.[7] Leigh Winsor, an expert on flatworms, believes that the "major factor" in transporting many invasive flatworms was the Wardian case.[8] In Winsor's taxonomic reappraisal of the shovelhead he described its occurrence in at least thirty-eight countries, from Canada to Costa Rica, from Israel to Ireland.[9] By 1899 the shovelhead was considered a "cosmopolitan" traveler, present in almost all the world's major botanical gardens and commercial greenhouses. For Winsor, the Wardian case "was the perfect way to transport planarians."[10] Among the places it was recorded was in the greenhouses at the botanical gardens in both Berlin and Paris. As Europeans colonized new lands, worms colonized greenhouses all over the world.

There are twelve nonnative species of flatworms in the United Kingdom, most of them from the former colonies of Australia and New

Zealand.[11] Flatworms are not capable of moving very far on their own: one study showed that in a year flatworms traveled only thirty meters.[12] Ecologists are becoming more and more concerned by the movement of flatworms. The New Guinea flatworm (*Platydemus manokwari*), for instance, has been found in France and the United States, where it has wiped out native snail populations. It is classed in the top one hundred of the world's most harmful invasive species. Scientists say the biggest disperser of these worms is the nursery trade.[13] The dispersal of flatworms has been going on for well over a century and has had major ecological impacts. Dispersals of soil play a major role in movement of flatworms around the world, and the impact of these movements continues to be felt today.[14]

In this final phase of its long and useful history, the Wardian case was used not only by botanists and nurserymen, but by entomologists, bureaucrats, and plant pathologists. While many species, such as the earthworm, were moved by accident, by the early twentieth century, with the emergence of programs of biological control, many insects were moved intentionally. The final significant journeys of the Wardian case were made for the purpose of moving not plants but insects. Indeed, many entomologists preferred to call them "Wardian cages." Larvae and parasites were moved so that they could feed upon, and hopefully destroy, invasive plants or insects. By moving plants we had meddled with a complex ecological world, but to control that world we adopted equally complex measures.

Pathogens and Plants

There are three basic types of plant pathogens: fungi, bacteria, and viruses.[15] Added to this are insect pests that threaten crops and biodiversity. At some time or another all of these were sent in Wardian cases. Of all the plant diseases circulated in the late nineteenth century, the devastation inflicted on the Ceylon (today Sri Lanka) coffee plantations was among the most well-known and also one of the earliest to demonstrate the need to restrict the free flow of plants.[16] British imperial expansion in the Arabian Sea, particularly in southern India and Ceylon, saw the expansion of large monocultures of coffee. Planters looking to make a greater return from the land introduced a West Indies style of planting that intensified land use and did away with shade trees. At its height Ceylon was the third largest coffee exporter in the world.

Coffee rust (caused by the fungus *Hemileia vastatrix*) was first recorded in Ceylon around the time of Ward's death, and the first epidemic broke out in Ceylon in 1869.[17] How the fungus first arrived in Ceylon is unknown, but the Wardian case played a key role in spreading the disease. After 1865 live coffee plants came to Ceylon from British Guyana, Cuba, Jamaica, Java and Liberia, any of which shipments could have brought the disease. The tropical, wet conditions in Ceylon and India perfectly met the needs of the fungus, and it spread rapidly. By 1870, although land under cultivation had increased, production had dropped dramatically. Losses were as much as £2 million per year. A similar destruction occurred in southern India plantations. In just two decades, by the 1880s the coffee industry in these regions had collapsed. The planters turned to other crops such as cinchona, cacao, rubber, and tea. Because plants were in constant circulation, coffee rust spread throughout much of Asia, Africa, and the Americas.

Interestingly, in June 1903 the disease was first spotted in Puerto Rico at the Mayaguez Experiment Station. The horticulturalist Otis Barrett noticed rust on small seedlings of Liberian coffee in a Wardian case sent from Java and quickly had the plants and the case destroyed.[18] The spread of the disease was halted temporarily, but eventually, in the 1970s, it arrived in Brazil and caused many problems as it spread throughout the Americas. Coffee rust is but one example of the threats that face agricultural monocultures, because when we focus on a single crop, and as it takes over more land, the risk of disease increases. The ecologist Rob Dunn tells it pretty clearly: economically planting just one crop is a simple way to turn a profit, but biologically it poses problems.[19]

There were other diseases. Root rot or water mold, caused by *Phytophthora*, ruins many plants. *Phytophthora* is part of a group that are very similar to fungi, the biggest difference being that as the disease takes hold of a host spores are released into the soil that allow the pathogen to travel in water, for example after rain, and spread to other plants in the vicinity. Generally, these are soil-based diseases and attack the roots of plants. *Phytophthora* can remain in the soil for many years waiting for conditions to be beneficial to travel and infect plants. Probably the most well-known plant disease outbreak in history—potato blight, which led to a famine that ravaged Europe and particularly Ireland from 1845 to 1852—was caused by *Phytophthora infestans*.[20] One agricultural disease expert suggests that potato blight was introduced to Europe in infected soil either in Wardian cases or in some earlier kind of traveling plant case.[21] Although the evidence for such a claim is thin, the timing fits.

Other species of *Phytophthora* had widespread impacts on avocado, banana, and citrus plantations. One of the most widely spread diseases in citrus is foot rot or collar rot. First recorded in the Azores in 1836, a stopping point along many transatlantic shipping routes, from the 1860s this variety of the disease spread to citrus-growing regions in Portugal, Italy, Australia, and the United States.[22] In Sicily it was said to have killed every tree on the island. Disease outbreaks were of great concern for Florida's flourishing citrus industry, so after root rot was first discovered there in 1879, restrictions on importations of citrus were tightened.[23] One variety, *Phytophthora cinnamomi*, has had a tremendous impact on avocado crops worldwide and also on native plants. These days, one way to avert root rot is by planting complementary plants together. In subtropical Australia, for example, bananas are often planted with avocados: the banana absorbs much of the water in the ground after heavy rain, thus helping the avocado. *P. cinnamomi* is a global problem. Scientists suggest that the importation of soil and living plants over the past 150 years caused the spread of the disease.[24] Today it is one of the major pathogens wiping out native flora in Australia's temperate forests.

The transport of rubber led to one of the stranger outcomes of moving plants. Not long after rubber plantations in Asia started producing rubber from the high-yielding *Hevea* variety, Brazil was hit by disease. The South American leaf blight caused by *Microcyclus ulei* devastated Brazilian crops from the start of the twentieth century (fig. 11.1). It was first discovered after the German explorer Ernst Ule sent a Wardian case of rubber plants to the Botanic Garden Berlin, many of them infested with leaf blight. Unlike the British shipment, this one included live plants,

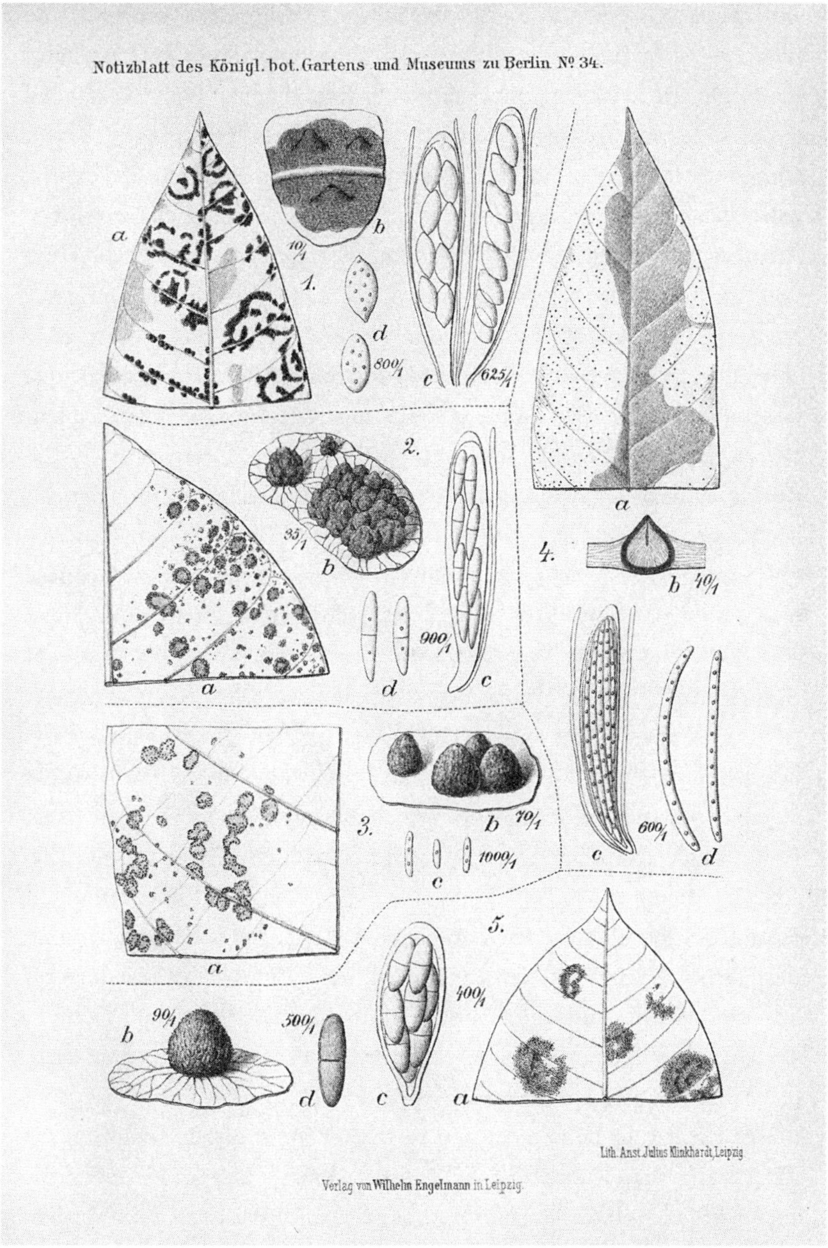
Notizblatt des Königl. bot. Gartens und Museums zu Berlin № 34.
Lith. Anst. Julius Klinkhardt, Leipzig
Verlag von Wilhelm Engelmann in Leipzig

FIGURE 11.1 South American leaf blight and other diseases affecting rubber plants, as illustrated by scientists from the Botanic Garden Berlin, 1904. From P. Hennings, "Über die auf *Hevea*-Arten bisher beobachteten parasitischen Pilze," *Notizblatt des königlichen botanischen Gartens und Museums zu Berlin* 4, no. 34 (1904): 139.

and plant pathologists in Berlin were able to identify the pathogen.[25] The fungus is actually native to Brazil, and for a long time the fungus and the *Hevea* tree lived together. But following its theft, when Brazil needed to compete with rubber flooding the market from British, French, and German colonial plantations, the country's plantings intensified from a native tree that was cultivated in small groves and interspersed in the Amazonian rainforest to a cleared landscape dedicated to a single crop. South American leaf blight spread rapidly, and Brazil's rubber industry has never recovered.[26] The blight has prevented large-scale commercial cultivation of rubber on the American continent. However, as the British had only stolen seeds, the fungus was not carried along for the ride. Plantations in the Asia-Pacific region, today the source of 90 percent of the world's natural rubber, so far have not been affected by South American leaf blight.

Many horticultural introductions have overrun landscapes. Japanese plants have not only made American gardens richer, they have spread widely. By 1918, just before Quarantine no. 37 went into effect, the Japanese honeysuckle (*Lonicera japonica*) had spread from Massachusetts to Florida, having been introduced in Wardian cases half a century earlier, in the 1860s, by George Rogers Hall with some of the first Japanese plants to arrive in New England (see chapter 7). Among the plants that Hall brought from Japan were Japanese chestnuts delivered to the Parsons & Co. nursery in New York. Hall's were certainly some of the first Japanese chestnuts to arrive in the United States. It is widely known that the chestnut blight pathogen was carried to the United States on nursery-imported Japanese chestnuts, but when it was imported is unknown.[27]

It is no coincidence that as the Wardian case became widely used, plant diseases also spread across the globe. It was also in the period of the Wardian case that we turned to greater focus on monoculture agricultural crops. Monocultures such as coffee, rubber, cacao, citrus, and sugar have all suffered from invasive plant diseases sent in Wardian cases. Surprisingly, in the twentieth century we did not turn away from monocultures but instead looked to quarantine and "natural enemies" to protect valuable monoculture crops. While plant pathologists focused on plant diseases, entomologists focused on pests that infested crops.

Natural Enemies

In the winter of 1888, the California entomologist Albert Koebele went to Australia looking for insects.[28] He traveled across much of the southern part of the continent looking for "natural enemies" of the cottony cushion scale (*Icerya purchasi*, also known at the time as fluted scale). This small insect ravaged crops across the globe, from South Africa to the Pacific Northwest. In the late nineteenth century it was described as an "insect which ranks amongst one of the most destructive pests injurious to plants."[29] It was assumed that it had arrived in California two decades earlier on shipments of wattle from Australia. It quickly moved to Californian citrus crops. Despite continuous efforts to stop the insect, it became such a problem in orange groves in California that "people find it impossible to keep it down."[30]

The entomologist Frazer Crawford in Adelaide had shown that the parasitic fly *Cryptochetum iceryae* fed upon the scale and could be a potential biological control. Small samples of the fly were sent to entomologists in the United States. This was Koebele's prime collection target. He traveled to Adelaide in search of the parasitic fly. On 15 October 1888 he visited citrus groves in the north of the city, where he not only found specimens of the fly but also had his first sighting of ladybugs (*Rodolia cardinalis*) feasting upon the scale insects.[31] Ladybugs are a diverse species with thousands of different varieties worldwide; this Australian variety had a particular appetite for the cottony cushion scale.

Koebele quickly arranged for infested citrus plants with their parasitic friends, the ladybugs and the flies, to be packed into cases and sent to California. To send insects he used wooden boxes and tin boxes, and "in addition, Dr. Schomburgh, director of the botanical gardens at Adelaide, kindly fitted up for me a Wardian case which was filled with living plants of oranges."[32] The twelve plants were infested with ladybugs. The case was handled in such a "rough manner" by the ship hands in Sydney that he thought no good would come of the shipment. Koebele noted that the Wardian case was a "bulky thing" that weighed more than 240 pounds.[33] Nonetheless, the insects survived the trip to Los Angeles.

There were enough ladybugs to infest a tree on a citrus estate outside

Los Angeles.[34] At F. W. Wolfskill's estate the Wardian and other cases were placed under an orange tree covered with a tent. The cases were opened and the ladybugs allowed to feed on the scale on the tree. In total Koebele sent three shipments with 12,000 samples of the fly and 514 ladybugs. The ladybugs proved superb predators and were widely distributed, and within a short time the tree on Wolfskill's property was free of scale. The tent was removed and the ladybugs set loose on the entire orchard, with results the same as on the single test tree. Other locations around California reported equal success. At one orchard the beetle had multiplied so much in four months that the owner, J. R. Dobbins, distributed over 63,000 ladybugs to growers in the area. Within just one year after the release, Los Angeles County had increased orange shipments from seven hundred to two thousand carload lots. The ladybug was spread throughout the world as an effective control against cottony cushion scale.

The biologist Paul de Bach, a leading proponent of biological control, has said that the success of the ladybug introduction "established the biological control method like a shot heard around the world."[35] It was the first and one of the most successful projects in biological control—aided by the Wardian case. In this late phase of its use, the case was employed to move entire ecosystems. Entomologists who witnessed the devastation of crops by insect pests turned to biological controls to contain them.

Over the next few decades, following the success of the ladybugs, the Wardian case became a key technology for moving parasitic insects. Further proof of the case's usefulness came in 1911, with Russell Woglum's transplants from India. Having gotten the idea of using Wardian cases from Charles Marlatt and David Fairchild, he filled six very large cases of infested live plants with specimens of the small parasitic wasp *Encarsia lahorensis*, a "natural enemy" of the whitefly.[36] In the introduction to his report on the project, the head of the Bureau of Entomology said that Woglum's use of the Wardian case had "demonstrated the correct methods" of shipping parasitic material.[37] In November 1911 he accompanied the cases from Bombay to New York, from where they were express shipped to Orlando, Florida. All the wasps arrived alive and healthy (fig. 11.2).

FIGURE 11.2 Six Wardian cases containing the natural enemies of the citrus whitefly leaving Lahore on their way to Orlando, Florida, 1911. Using Wardian cases for this transport was suggested by both David Fairchild and Charles Marlatt. Below are the native Hindustani men who collected the insects. From Russell Woglum, *A Report of a Trip to India and the Orient in Search of the Natural Enemies of the Citrus White Fly* (Washington, DC: Government Printer, 1913).

Another interesting use of the case came in 1929, when Cuban and American scientists collaborated in finding a biological control for the blackfly, coincidentally named after Russell Woglum, *Aleurocanthus woglumi*.[38] Entomologists traveled to Singapore and Malaysia, the black-fly's native home, where they identified at least fourteen insects that fed on the pest. But the main predators that kept the fly in check were internal parasites, so the American entomologists imported the para-site *Eretmocerus serius* to Cuba, with mixed results. But again, the chief technology used to move the parasites and other natural enemies was the Wardian case. The American entomologists adopted the design of Wardian cases from the cases used by the Singapore Botanic Gardens, so they were "of the type employed by the Department of Agriculture of the Federated Malay States in their plant shipments to England."[39] In total, three shipments were sent from Malaysia to the United States and on to Cuba.

The collaboration of scientists across national borders is an inter-esting global current running through much of this work of biological control. American and Australian scientists collaborated on the cottony cushion scale in the 1880s, and Cuba and the United States worked together on the blackfly, for which they adopted Wardian case designs from Singapore.

Caged Moths

In 1925 the Wardian case was again put to use in what is championed today as one of the most successful uses of insects for biological control. This time the American moth *Cactoblastis cactorum* was moved to Aus-tralia in an attempt to control the invasive prickly pear cactus. Originally from the arid regions of Mexico and the southern United States, the hardy prickly pear (*Opuntia* spp.) arrived in Australia with the first fleet of British colonists in 1788. By the early nineteenth century other variet-ies of the prickly pear (particularly *Opuntia inermis* and *Opuntia strictus*) had been introduced by the nursery industry. It was to serve as a fod-der plant for livestock, as a garden ornamental, and as an edible fruit.[40]

Today the fruit is still consumed in Mexico, and the plant also serves an important part of the pastoral economy in Madagascar.

Following its introduction in Australia and South Africa, the cactus spread rapidly, choking useful farmland. The director of the botanical gardens in Cape Town, Peter MacOwan, described the problem as early as 1888: "The land occupied by them is practically useless for all purposes of husbandry."[41] Even after the concerted efforts of botanists and the offer of rewards for technological and chemical solutions to the problem, it continued for decades. In Australia, at its height in 1925, the prickly pear cactus infected more than twenty-four million hectares.[42] Many farmers in northern Australia simply walked away from their land.

Many attempts were made to control the cactus. State governments in Australia formed prickly pear control boards to investigate the solution. In 1912 they sent scientists on an eighteen-month world tour in an attempt to find some way to control this invasive species. The main recommendation that emerged from this research trip was that a biological solution needed to be found.[43]

Many different insects were shipped from the Americas to Australia for testing. There were repeated failures, with many insects dying in transit owing to shipping delays and poor transportation techniques. The American John Hamlin was hired in 1920 by the Australians and brought his knowledge from the USDA's Bureau of Entomology to Australian prickly pear research. His biggest contribution was to modify the Wardian case so that it could successfully ship insects.[44] Hamlin certainly had firsthand experience with the Wardian case: his first job after graduating from Ohio State University was as a plant quarantine inspector for the Federal Horticultural Board. Under Hamlin's instruction, fine mesh was inserted on both sides of the sloping roof, and the case was protected on the outside by strong wire and iron strips. Sphagnum moss was placed in the bottom of the cases, which were then filled with prickly pear plants. The parasitic insects were attached to the plants. Another addition that the entomologists made was to construct a flat piece of wood across the apex of the cases so that they could be stacked but light and air could still enter.

FIGURE 11.3 Wardian cases full of insects ready for shipment from Uvalde, Texas, to Australia, ca. 1925. From Alan Dodd, *The Progress of Biological Control of Prickly-Pear in Australia* (Brisbane: Government Printer, 1929).

In 1920 Australia invested heavily in a coordinated national program called the Commonwealth Prickly Pear Board. By this time it was known that the native home of the prickly pear was Central America. The Australians worked closely with their American colleagues at the USDA's Texas station.[45] Led by the Australian entomologist Alan P. Dodd, from there they scoured the United States, Mexico, Guatemala, El Salvador, Honduras, the West Indies, Panama, Colombia, Ecuador, Peru, Venezuela, Brazil, Uruguay, and Argentina. Dodd wrote of the insect imports: "Consignments of insects from North and South America have, from the commencement of the investigation, been made in specially constructed Wardian cases" (fig. 11.3).[46] Within a decade, Dodd reported that they had made forty-five shipments from the Americas comprising 566 cases and "many thousands of insects of various kinds."[47] This was a large number of cases—at least seventy per year. Consider that by the turn of the century, botanical gardens such as Kew and the Jardin d'Essai Colonial were shipping between twenty and forty cases per year.

In late 1924 the *Cactoblastis* moth was observed in Argentina, near Concordia. Dodd waited until early 1925 for the *Cactoblastis* caterpillars to turn to moths and lay eggs. Three thousand eggs were collected and carefully deposited on prickly pear plants that were waiting in the specially constructed Wardian cases. Loaded in Concordia, the cases were shipped to Buenos Aires, and from there they were ten weeks at sea.

First the boat stopped at Cape Town, and colleagues at the South African Department of Agriculture took out 250 larvae for their own trials. The cases were then loaded onto the first boat for Sydney and from there sent by rail to Brisbane. At the biological test station they quickly discovered that the insects thrived on the prickly pear (both *O. inermis* and *O. stricta*). Within the first two seasons at the test stations, more than two million eggs had been produced from Dodd's first shipment. The moths were liberated across a number of sites in northern Australia. The results were successful.

Over the coming years millions of eggs were bred and distributed to infested areas, with the same results. The work was intensive. Landholders and workers had to individually affix the eggs onto squares of paper and then attach these to the cactus pads; when the eggs hatched, they tunneled into the pad, fed, and destroyed the plants. Within seven years of the Wardian cases' arrival in Brisbane the prickly pear was destroyed. There were further releases to control regrowth over the coming years, but ultimately, within a decade of the release of the *Cactoblastis*, the prickly pear was no longer a problem in Australia.

Surprisingly, following the success of the *Cactoblastis*, the importation of foreign insects did not stop but increased. From 1927 to 1930, just as the *Cactoblastis* was ravaging the prickly pear in Australia, there was an increase in the number of insects sent to Australia. In one season alone, 1928–29, the Australian and American entomologists sent 375 Wardian cases. Three hundred and seventy-five. Clearly the entomologists were buoyed by their success and thought it necessary to keep experimenting with insects. Over fifteen years they made a total of sixty-seven shipments to Australia. Over 500,000 insects of forty-eight different species were sent in a total of 1,230 Wardian cases. The use of the Wardian case by the Prickly Pear Board was some of the most intensive use in its history.[48]

Last Journeys

The final uses of Wardian cases for biological control were in 1948, with consignments of further parasites for citrus blackfly. But for that shipment air freight was also used, which became the more efficient method.[49] Although biological control is still used today, by the 1950s synthetic organic insecticides had become relatively cheap and were being widely used.

There were many other insect transfers using the Wardian case. The case was used to import the coconut moth parasite *Bessa remota* from Malaya to Fiji. To some this was successful, but it had dire consequences. Unlike other pests that entomologists were trying to get rid of, the coconut moth was unique, beautiful, and endemic to Fiji. When they wiped it out, they brought about the extinction of the species. This work was conducted by the Imperial Bureau of Entomology, the British equivalent of the American bureau.[50] Biological control is a controversial method, and there are many well-known cases of its triggering further pests or having unintended consequences. Indeed, today the *Cactoblastis* moth has now found its way into the United States and the Caribbean and is threatening many native cactus species.

In its last phase, the Wardian case was used to send plants infested with insect parasites. The irony of this long history should not be lost on us. In just a century of activity, people moved large numbers of plants, with important consequences; but at the same time, we unleashed a whole range of complex ecological relationships over which we had little control. In the final act we again turned to the Wardian case to help us. Control of nature allowed us to move plants around the world, but nature's persistence shows us that our control was always conditional.

Conclusion
....

Case Closed?

After a short walk from the suburban train station, I entered Nogent-sur-Marne, Paris's largest public park on a crisp December morning. There were joggers and dog walkers, the yellow and brown leaves from the tall trees shattered underfoot, the sky was blue, and the winter half-light made the scene look all the more unreal. But this was no ordinary park. It was once the Jardin d'Essai Colonial and the host of the Exposition Coloniale in 1907. That year more than two million people came to the park to see the products and the symbol of empire. Once it was a thriving garden and symbol of empire; today it is a decaying ruin of an era often forgotten.

Many of the buildings and greenhouses are still there. The foreign pavilion brought from French Guiana is boarded up and weathering. Like many other buildings, the Madagascar pavilion has long since collapsed. The Vietnamese building still stands and retains much of its former splendor, maintained as part of the research facility that still operates on the grounds. As for the greenhouse, where so many Wardian cases were once packed, many of the glass panes are smashed, and the inside has been overrun with weeds. Hundreds, probably thousands, of Wardian cases passed through here in the late nineteenth and early twentieth centuries. But today there is a strange silence where past meets present. World-renowned botanical gardens such as Kew and the Jardin des Plantes and the Botanic Garden Berlin have been maintained, upgraded, and transformed into sites of biodiversity and conservation. The ruins of the Jardin Colonial are a reminder of the former world of moving plants—of the close relationship between colonial gardens that sent plants out and of a broader sweep of colonial practices that emanated from similar sites and nations.

Today the gardens are part of the French Agricultural Research Centre for International Development (CIRAD), which does important work for the sustainable development of tropical regions, mostly in developing countries. At the site in Nogent-sur-Marne there is a library and archive that preserve the history of the Jardin d'Essai Colonial—and of the Wardian case. I spent my day in the archive following the fragmented tale of Wardian cases as they went around the world. After I finished working with the old documents, I walked along the garden's paths. As dusk faded into night I kept circling back to the old greenhouse. So many of the most captivating images of the Wardian case were taken outside the greenhouse; but now it was falling down (fig. C.1). I was reminded how easy it is to forget past practices unless they are made visible.

The practice of using Wardian cases to move plants has largely been forgotten. A few months after visiting the gardens in Paris, I was in the American Midwest, driving across the prairies, trying to concentrate on the road without being distracted by the epic rise of the Rocky Mountains ahead. I was visiting the National Center for Genetic Resources Preservation, in Fort Collins, Colorado. It not only holds a rare full set of the *Inventory of Seeds and Plants Imported* from 1898 to 1942 that David Fairchild was so influential in starting, but, more important, is the site of the USDA's seed bank for the preservation of germplasm (plant material). For the United States, this is where the genetic material of plants and animals are stored in case we might need them—for the future of agriculture, for the future of our survival.

I was taken through the whole process, from seed sorting to storing to cryogenic preservation. I saw how seeds and plant material arrive in simple US Postal Service boxes and are then sorted and tested by trained scientists. Those seeds that are viable and unique are kept and stored. On

FIGURE C.1 Greenhouses in ruins at the Jardin d'Essai Colonial, Nogent-sur-Marne, Paris. Photo by the author.

this day, the seeds they were testing were not responding well. "There is no point in keeping the seeds if they are not germinating," the sorting scientist tells me. They are arid plants that are being collected in an effort to preserve the native flora of the American West. After a third attempt the analyst has managed to get the seeds to germinate, so they will be preserved. Seeds and germplasm are collected from all over the globe. When they are ready for preservation, they are taken to a large storage room that is kept at a constant −18 degrees Celsius. Seeds can remain under these conditions for over a hundred years (legumes might even keep for four hundred years) and still be taken out and grown.

From the seed storage area we take the elevator down two levels to the underground vault, a specially constructed storehouse for nature's richest treasures. Capable of withstanding fire and earthquake, it is an underground ark of sorts. Inside the vault the plant material, fragments of plants that will allow scientists to collect genetic material for breeding, are stored in large tanks of liquid nitrogen, which preserves the genetic quality of the material. Next to the plant material are similar tanks full of animal embryos and semen, even microbes. Protecting and preserving the diversity of plants and animals, which is vital to human life, entails sealing it in bank-style vaults and keeping it secure for hundreds of years. They are preserved as a backup in case our most relied-upon agricultural crops become so devastated by diseases or pathogens that we need to breed new material. Things have changed a great deal from the days when useful plants were sent across the oceans. Today new plants are more likely to come from a scientist wearing a lab coat than an explorer with dirt and mud on his or her boots.

Some of the plants that need protecting and possible engineering are ones that the Wardian case moved: major agricultural commodities such as avocados, bananas, cacao, coffee, mangoes, rubber, and tea. Every one of these crops is grown in large-scale monoculture plantations, many of them not in the plants' native homes. Cacao, the primary ingredient in chocolate, is one crop that was sent from the New World to overseas colonies in Wardian cases. Today the Ivory Coast, Ghana, and Cameroon—former French, British, and German colonies respectively—produce most of the world's cacao. Although the way we use and manipulate plants is very different from the techniques and purposes of Nathaniel Ward, there is a strong connection between the plant products we most value today and the history of moving plants in Wardian cases.

Last Journeys

The use of the Wardian case by entomologists for biological control was its final large-scale effort. Beginning in the 1930s, the case's use waned, until it finally made its last journeys in the 1960s. Over these three decades there were smaller shipments of cases, but their impact and value was superseded by other technologies. A number of factors—environmental, technological, even cultural—contributed to the cases' decline. As the previous two chapters show, the spread of invasive species and pathogens was a significant turning point, not only in the long history of the Wardian case, but more generally in our relationship with the environment.

This century of use, from the first journey in 1833 to the final significant ones in the 1930s, saw a dramatic change in the ways that we think
about moving plants. At first there was widespread enthusiasm about
a technology that could move plants successfully. Over the following
century, however, we came to see that all of this movement had consequences. In 1945 the USDA advised that plants not be shipped in soil:
not only was it very expensive, but most countries banned the importation of soil because of the high risk that it could "contain dangerous
insects or diseases."[1] By the 1940s, with increasing quarantine restrictions
on soil in particular, the Wardian case was being phased out.

There was also technological change. By the 1930s, sending plants
by air transport dramatically shortened travel times. In 1931 David Fairchild sent freshly collected seeds and cuttings from the Caribbean by
air freight; they took only five days to arrive at the quarantine house in
Washington. There were even some reports of Wardian cases traveling
by air freight. With shorter travel times, keeping plants alive on journeys
became much easier. The Wardian case was superseded by the use of
polythene bags and temperature-controlled air transport. Today horticulturalists who send plant material will brush off all the soil, trim the
plant back until it is almost a stick with roots, wrap it in a sealed plastic
bag, and box it up for the courier company to transport it to the next
location. If it is an international shipment, then an import and export
license will have to be attached to the boxes and it will need to pass
through a quarantine station.

Other factors also altered the trade in live plants. An important one
was changing tastes: there was a shift away from greenhouse varieties,
and hardy plants became more desirable. In some places this meant
plants that could survive harsh winters. In Boston many hardy Japanese
and Chinese varieties from similar climates became popular, while farther to the south palms were the flavor of the day. Many hardy varieties
could be transported as seed. Preferences, tastes, and fashions are always
changing: these days, with more people living in small apartments and
having limited time and gardening space, the terrarium has undergone a
resurgence to become a chic feature of the modern home.[2]

Although Wardian cases fell out of favor throughout the twentieth century, Kew still remained famous for them and was regularly called upon to provide information about them. As late as 1950 an article in *National Geographic* on the beauty and scientific knowledge fostered by Kew was not complete without showcasing "plant life's traveling greenhouse."[3] The last journey of a Wardian case to Kew was in 1962, when a box of ornamental plants arrived from Fiji. Kew Gardens also helped to memorialize the case. In 1987 they worked with the Morris Arboretum at the University of Pennsylvania on the exhibition *Plants under Glass*, in which the Wardian case was featured. The exhibition was shown at both the Pennsylvania Horticultural Society's Philadelphia Show and the Royal Horticultural Society's Chelsea Flower Show. For the exhibition Kew packed a Wardian case full of endangered plants and sent it over the Atlantic by ship. From exotic ornamentals to endangered species, from sitting rooms to flower shows, from economic plants to exhibitions, moving plants had changed a great deal by the end of the twentieth century.

Today botanical gardens and arboretums have been rebranded as storehouses of biodiversity and centers of environmental knowledge. By some estimates there are nearly two thousand botanical gardens worldwide, and more than 250 million people visit them each year. And they are growing in popularity.[4] As cities become denser and people search out green space, gardens offer many benefits, including exercise, knowledge, and restoration. If the garden you are visiting dates back more than a century, then many of the plants growing in them traveled a great distance before their roots were finally placed into the soil. And if the color and diversity on show in one of those gardens is impressive, then it would be good to remember that in the previous century the availability of foreign plants was possibly greater than it is today.

The Lag Effect

Near the end of this project, after I had returned home to Australia, I was out on a day hike in the bush near my home when I discovered that many of the impacts of the Wardian case are also very local. I was walking with my sons and daughter through bush on the Victorian surf coast, a wild and rugged part of southeastern Australia. In 1854, on the beach not far from here, Ward's close friend William Henry Harvey collected hundreds of unique specimens of algae, sending many of the duplicates to London for Ward's microscope parties.

Our path was lined with grass trees (*Xanthorrhoea*), a distinctive Australian plant with a long trunk and hundreds of protruding green spiky leaves that look like either a patch of long grass or a bad hair day. When in flower a straight spear, sometimes up to four meters long, grows out of the nest of spikes and produces a spiral of nectar-filled flowers. The trees grow very slowly and are fire resistant. My children knew the plants from a widely read children's-book series, *Grug*, by Ted Prior, in which one of these trees (named Grug) does everything from building a car to becoming a superhero. Walking past this grove of plants, my children were concerned. One of my sons asked, "Papa, what is wrong with Grug?" Some of the trees had lost their green spiky top, on others the green foliage was turning black, and still others were just trunks, uprooted and lying on the ground.

Grass trees in southeastern Australia are under threat from the plant pathogen *Phytophthora cinnamomi*. Dieback, as it is commonly called, is devastating many native plants throughout Australia. By some reports more than 2,500 species of Australian plants are affected by the disease.[5] It arrived on infected soil with settlers in the nineteenth century, most likely in Wardian cases, but did not become a major problem until the 1960s, when it devastated jarrah trees in Western Australia. More than 70 percent of one national park in Western Australia is infected. It has since spread all across the continent. Invasion biologists call this time between when an invasive species arrives in a country and when it commences taking over the native population and causing problems *lag time*.[6]

I told my son the truth: that many years ago we had brought the virus and now it was a problem. That these plants would die because their roots were rotten, and there was no cure. He nodded, in some ways understanding the starkness of life and death, and we kept walking. Along the walk we noticed more dying trees. The grass tree is just one localized example of the problems caused by the movement of plants. You probably have only to look outside to see other examples in your own environment.

For too long the Wardian case has occupied a heroic space in horticultural and botanical folklore—myths about a doctor who invented a box that changed the world. They do not tell the full story. This book has made a long journey, salvaging fragments of the historical record to unearth the hundred-year history of the Wardian case. As important as Nathaniel Ward's idea of an enclosed system was, his excellent connections to scientists and gardeners in natural history circles in London and Europe were just as crucial, since these prominent individuals helped to spread word of his case. Ward's house, we remember from chapter 6, "was the most frequented metropolitan resort of naturalists from all quarters of the globe of any since Sir Joseph Banks' day." Such friendship and collegiality allowed for the quick uptake of the case and for Ward's name to be forever attached to it.

The value of this technology went well beyond Ward or even Victorian London. Dutch, French, German, and American imperialists used the case to great effect. Nor can we discredit the widespread use of the case by nurseries and gardeners from all over—Exeter, Cape Town, Brussels, Rio de Janeiro, New York, Fukuyama. As the Wardian cage shows, the single biggest use of the case in any one year occurred in the twentieth century, with Australian entomologists moving insects from the Americas to Australia. The Wardian case helped transform the global environment.

The Royal Botanic Gardens, Kew, has featured prominently in this story, in part because Kew has a very good archive. But as I have tried to show throughout, it was only one of many important sites that used the Wardian case. The United States, France, Germany, and Australia were key users, but so were Brazil, Bangladesh, China, India, Japan, Madagascar, Samoa, and Sri Lanka, to name just a few locales. In the latter group there is still more work to be done to dig up details regarding the local experts and workers who toiled so hard to help colonists and imperial scientists identify and send plants.

The Wardian case provided possibilities for moving plants. New technologies provide the lure of possibility. And it was a powerful tool in the hands of boosters, promoters, and imperialists. Particularly in the mid-nineteenth century, knowing that plants could be moved at all was an important motivation for moving them in the first place. The Wardian case was a widely promoted method that connected plants and people over long distances. It was also an idea that told gardeners, colonists, scientists, and bureaucrats that plants could be moved. Possibility also has a symbolic affect. By the late nineteenth century, new imperialists could promote the case to show their work of spreading agriculture to the colonies. With possibility on its side, the case was used widely.

Over its long life, the Wardian case as a witness saw much, but as a carrier it brought a lot more than plants. By its very design, it was a moving ecosystem, which means that it brought with it species and diseases that wreaked ecological havoc. Today it is recognized that nonnative species, including many plants, are one of the major drivers of global biodiversity loss. The movement of plants also created huge economic impacts that have been calculated in the billions of dollars. Weeds — destructive plant species — have been estimated to cost the Australian economy as much as $3 billion and the US economy more than $26 billion annually.[7] To be sure, the Wardian case was not the only carrier; but as the prime mover of plants it had a hand in bringing about significant ecological changes, many of which we are still trying to come to grips with.

An environmental critique of the Wardian case needs to be tempered with an acknowledgment of its importance in moving valuable plants. Avocados, bananas, cacao, coffee, cinchona, fruit trees, timber trees, mangoes, rubber, and tea were some of the key species. There were also ornamental plants: daphnes, fuchsias, ferns, ixora, magnolias, rhododendrons, and roses. In settler societies, many of the foreign ornamentals that now enrich suburban gardens arrived in Wardian cases.

The nursery trade in the nineteenth century was extensive and global and was intimately connected to botanical institutions. Often expeditions in search of plants were sponsored by state institutions and commercial nurseries, a practice that continued well into the twentieth century. Nurseries such as George Loddiges, Harry Veitch, James Backhouse, William Bull, and Hugh Low used the case extensively, shaped its design, and encouraged others to adopt it. Commercial nurseries were as important in moving nature as were imperial scientific institutions.

Many hands worked with the Wardian case. There were plant hunters, local collectors, indentured laborers, indigenous peoples, gardeners, curators, botanists, entomologists, horticulturalists, amateurs, carpenters, diplomats, men, women, and children. These were the people who built, packed, sent, and accompanied the cases, many of which were constructed by skilled carpenters in cities such as—Hobart, Lima, Rio de Janeiro, Tokyo. In the last instance, so good was the work of the Japanese carpenters that Kew reused their cases to send important plants to colonial botanical gardens in the Caribbean.

I began this journey after my imagination was captivated by the Wardian case and I began to wonder why so few of these important objects remained in museum collections. The answer was quite simple: most of them were probably burned because of quarantine restrictions. In this simple answer lies a much grander story of how our environmental practices have changed over the last two centuries—from moving plants to restricting their flow. But the balance of nature never performs to our human wishes. We are still dealing with the consequences of moving plants in Wardian cases.

Acknowledgments

This book was made possible through the generous financial support of a number of institutions and organizations. I am deeply grateful to the Gerda Henkel Foundation, Dusseldorf, for their research scholarship; the Deutsches Museum, Munich, and the Kulturstiftung des Bundes (Federal Cultural Foundation), Halle, for the FellowMe curatorial fellowship; the Arnold Arboretum of Harvard University, Boston, for the Sargent Award (and for providing a home institution for much of this project); the University of Melbourne and the State Library Victoria, Melbourne, for the Redmond Barry Fellowship; the Huntington Library and Botanical Gardens, San Marino, California; and the Institute for Art History, Max Planck Society, Florence, and the Prussian Cultural Heritage Foundation, Berlin, for the 4A Lab Fellowship Program. I extend my deepest thanks to all of the people at each of these places who believed in the project and helped out along the way. I also wish to thank the Australian Academy of the Humanities for their financial support, through their Publication Subsidy Scheme, in producing the color plates in this book.

My study of the Wardian case took me far afield, and a number of institutions offered in-kind support for the project. Thank you to the Rachel Carson Center for Environment and Society, Ludwig Maximilians University, Munich; the Center of the American West, University of Colorado, Boulder; and the School of Life and Environmental Sciences, Deakin University, Geelong, Australia.

A topic as fragmented as the Wardian case must rely on archives and libraries from far and wide. I am grateful to the many institutions and people that helped me to track down documents and images and made them available for research and publication. In the United Kingdom, thank you to the Archives and the Economic Botany Collection at the Royal Botanic Gardens, Kew; Lindley Library, Royal Horticultural Society of London; the Archives of the Worshipful Society of Apothecaries, London; and the Kensington and Chelsea Local Studies Library, London. In Europe, thank you to the Museum of Natural History, Paris; the CIRAD Archives, Paris; the Botanic Garden and Botanical Museum Berlin; the Bavarian State Library, Munich; and the Munich Botanical Garden. In the United States, thank you to the Archives of American Gardens, Smithsonian Gardens, Washington, DC; the Archives of the Arnold Arboretum and the Gray Herbarium Library, Harvard University, Boston; the Missouri Botanical Garden Archives, St. Louis; the Mertz Library, New York Botanical Garden; the National Agricultural Library, USDA, Beltsville, Maryland; and the Library and Archives of the Fairchild Tropical Botanic Garden, Coral Gables, Florida. A special thank-you to the National Laboratory for Genetic Resource Preservation, Fort Collins, Colorado, for a wonderful tour of their facility. In Australia, thank you to the Library of the Royal Botanic Gardens, Melbourne; the Daniel Solander Library, Royal Botanic Garden, Sydney; the State Library Victoria, Melbourne; the State Library of South Australia, Adelaide; the National Library of Australia, Canberra; the Alfred Deakin Prime Ministerial Library, Deakin University, Geelong; the University of Melbourne Archives; and the Waroona Historical Society and Museum, Western Australia. Also, thank you to the online Biodiversity Heritage Library.

Thank you to the team at the University of Chicago Press, in particular Rachel Kelly Unger and Michaela Rae Luckey, for their enthusiasm and for steering the book to publication; and a special thanks to Christine Schwab, Nicholas Lilly, and Barbara Norton. Also, thank you to Kew Publishing for their support of the book.

Many people have helped me along the way. Thank you to Hannah Baader, Padraic Fisher, Jackie Kerin, Patty Limerick, Christoph Mauch, Nina Moellers, Cameron Muir, Celmara Pocock, Libby Robin, Dan Rosendahl, Susie Shears, Peter Spearritt, and Paul Sutter. A very special thank-you to Mark Nesbitt at the Royal Botanic Gardens, Kew, and Helmuth Trischler at the Deutsches Museum, Munich, for their enthusiasm and support for the entire project. To Marion Stell, thank you for your friendship and support over many years. A big thank-you to Julie Keogh and Robert Keogh for helping me out from time to time, and especially for your many visits to all those places where I found myself.

Finally, and most important, thank you to my family, who accompanied me on this journey. In order to allow me to complete this book, we lived on three different continents and in three times as many houses. Together we have experienced all the challenges that go with uprooting. But just like new growth on a plant in spring, every morning you were all there with new and beautiful surprises. To Hugo, Leopold, Emil, and Sophia, thank you for your courage, drawings, patience, and smiles. Finally, to Angela Kreutz, thank you for being the best traveling companion anyone could ever wish for and for everything else. I am looking forward to our next adventure together.

Notes

....

Introduction

1. Charles Mallard to Ward, 23 November 1833, Directors Correspondence (hereafter cited as DC), Archives of the Royal Botanic Gardens, Kew (hereafter cited as RBGK), vol. 8, 153.

2. John Livingstone, "Observations on the Difficulties Which Have Existed in the Transportation of Plants from China to England, and Suggestions for Obviating Them," *Transactions of the Horticultural Society of London* 3 (1822): 421–29.

3. Robert Friedel, *A Culture of Improvement: Technology and the Western Millennium* (Cambridge, MA: MIT Press, 2007).

4. This is not a book about indoor gardening or the fern craze; these topics have been covered extensively elsewhere. See, for example, David Allen, *The Victorian Fern Craze: A History of Pteridomania* (London: Hutchinson, 1969), as well as Sarah Whittingham's excellent *Fern Fever: The Story of Pteridomania* (London: Francis Lincoln, 2012), which covers the topic thoroughly and also notes the "important legacy" of the traveling Wardian case (19). On the orchid craze, see Jim Endersby, *Orchid: A Cultural History* (Chicago: University of Chicago Press, 2016).

5. J. D. Hooker to James Hector, 14 December 1884, in *My Dear Hector: Letters from Joseph Dalton Hooker to James Hector, 1862–1893*, ed. John Yaldwyn and Juliet Hobbs (Wellington: Museum of New Zealand, 1998), 181.

6. Although their lines of inquiry are quite different, I am inspired here by the "use-based history" of Ruth Oldenziel and Mikael Hard, *Consumers, Tinkerers and Rebels: The People Who Shaped Europe* (Basingstoke: Palgrave, 2013); and of David Edgerton, *The Shock of the Old: Technology and Global History Since 1900* (London: Profile, 2006).

By taking a longer and global view of the story of the Wardian case, we begin to see it as one of use and practice.

7. Phillip J. Pauly, *Fruits and Plains: The Horticultural Transformation of America* (Cambridge, MA: Harvard University Press, 2008), 1.

8. Civil society profoundly shaped global plant movements in the nineteenth century. See Patrick Manning, "Introduction," in *Global Scientific Practice in an Age of Revolutions, 1750–1850*, ed. Patrick Manning and Daniel Rood (Pittsburgh: University of Pittsburgh Press, 2016), 1–18, at 11–12. Manning is reflecting on Stuart McCook's essay in the same volume, "'Squares of Tropic Summer': The Wardian Case, Victorian Horticulture, and the Logistics of Global Plant Transfers, 1770–1910," 199–215.

9. Mark Levinson, *The Box: How the Shipping Container Made the World Smaller and the Global Economy Bigger* (Princeton, NJ: Princeton University Press, 2006). Another interesting history of boxes, of the few available, is Anke te Heesen, *The World in a Box: The Story of an Eighteenth-Century Picture Encyclopedia* (Chicago: University of Chicago Press, 2002). As for the phrase "prime mover," I am borrowing from Vaclav Smil; see Smil, "Two Prime Movers of Globalization: History and the Impact of Diesel Engines and Gas Turbines," *Journal of Global History* 2, no. 3 (2007): 373–94. See also Luke Keogh, "The Wardian Case: Environmental Histories of a Box for Moving Plants," *Environment and History* 25, no. 1 (2019): 219–44.

10. I am drawing here on Lynn Hunt, *Writing History in the Global Era* (New York: W. W. Norton, 2014), 62–77.

11. Many of these contemporary practices were born in 1951 with the United Nations' first International Plant Protection Convention. The Convention was largely a response to the open trade in plants in the previous century; see Andrew M. Liebhold and Robert L. Griffin, "The Legacy of Charles Marlatt and Efforts to Limit Plant Pest Invasions," *American Entomologist* 62, no. 4 (2016): 218–27.

12. Looking at the history of people moving plants is a well tilled field; key examples include Ray Desmond, "Technical Problems in Transporting Living Plants in the Age of Sail," *Canadian Horticultural History* 1, no. 2 (1986): 74–90; Marianne Klemun, "Globaler Pflanzentransfer und seine Transferinstanzen als Kultur-, Wissens- und Wissenschaftstransfer der frühen Neuzeit," *Berichte zur Wissenschaftsgeschichte* 29, no. 3 (2006): 205–23; Marianne Klemun, "Live Plants on the Way," *Journal of the History of Science and Technology* 5 (2012): 30–48; Donal P. McCracken, *Gardens of Empire: Botanical Institutions of the Victorian British Empire* (London: Leicester University Press, 1997); and Nigel Rigby, "The Politics and Pragmatics of Seaborne Plant Transportation, 1769–1805," in *Science and Exploration in the Pacific: European Voyages to the Southern Oceans in the Eighteenth Century*, ed. Margarette Lincoln (Suffolk: Boydell Press, 1998), 81–102. Alfred Crosby's pioneering book *Ecological Imperialism* looms large for any historian delving into plant transfers in the colonial period; Crosby, *Ecological Imperialism: The Biological Expansion of Europe, 900–1900* (Cambridge: Cambridge University Press, 1986). See also Eric Pawson, "Plants, Mobilities and Landscapes: Environmental Histories of Botanical Exchange," *Geography Compass* 2, no. 5 (2008): 1464–77, at 1474.

And this story not been told by historians alone. In Charles S. Elton's classic work *The Ecology of Invasions by Animals and Plants* (London: Methuen, 1958), which itself formed a foundational work for the science of invasion biology, he describes how plants and animals traveled as part of the baggage of humans moving around the globe. See also David M. Richardson, *Fifty Years of Invasion Ecology: The Legacy of Charles Elton* (West Sussex: Wiley-Blackwell, 2011); and Libby Robin, "Resilience in the Anthropocene: A Biography," in *Rethinking Invasion Ecologies from the Environmental Humanities,* ed. Jodi Frawley and Iain McCalman (New York: Routledge, 2014), 45–63.

13. Military historians show that the simple act of either supplying goods or cutting off their supply has profoundly impacted warfare since the late eighteenth century. The supply of ornamental plants to the colonies had an intimate relationship with botanical science and understanding native plants in imperial centers. The classic in military history is M. van Crevald, *Supplying War: Logistics from Wallenstein to Patton* (Cambridge: Cambridge University Press, 1977). Although not a history, the most recent and exciting contribution to this discussion is by the landscape architects Pierre Bélanger and Alexander Arroyo, *Ecologies of Power: Countermapping the Logistical Landscapes and Military Geographies of the U.S. Department of Defense* (Cambridge, MA: MIT Press, 2016). On the new interest by critical theorists in logistics, see Deborah Cowen, *The Deadly Life of Logistics: Mapping Violence in Global Trade* (Minneapolis: University of Minnesota Press, 2014); and Brett Neilson, "Five Theses on Understanding Logistics as Power," *Distinktion: Scandinavian Journal of Social Theory* 13, no. 3 (2012): 323–40.

14. On the exhibition, see Nina Moellers, Christian Schwaegerl, and Helmuth Trischler, eds., *Welcome to the Anthropocene: The Earth in Our Hands* (Munich: Deutsches Museum, 2015); and Libby Robin, Dag Avango, Luke Keogh, Nina Möllers, Bernd Scherer, and Helmuth Trischler, "Three Galleries of the Anthropocene," *Anthropocene Review* 1, no. 3 (2014): 207–24. For an excellent overview of the concept see Helmuth Trischler, "The Anthropocene: A Challenge for the History of Science, Technology, and the Environment," *NTM* 24 (2016): 309–35.

15. David Beerling, *The Emerald Planet: How Plants Changed Earth's History* (Oxford: Oxford University Press, 2007). On vascular plant dispersal, see Richard N. Mack and Mark W. Lonsdale, "Humans as Global Plant Dispersers: Getting More Than We Bargained For," *BioScience* 51, no. 2 (2001): 95–102.

16. I have carried out extensive correspondence with many institutions across the globe to find Wardian cases. Information on Wardian cases in each place listed is from personal correspondence with curators and employees. There are reports of a few more cases that have been elusive, possibly one in the botany department of the University of Auckland and another at the Missouri Botanical Garden; these have not been confirmed.

17. Jennifer Newell, Libby Robin, and Kirsten Wehner, "Introduction: Curating Connections in a Climate-Changed World," in *Curating the Future: Museums, Communities and Climate Change,* ed. Jennifer Newell, Libby Robin, and Kirsten Wehner (London: Routledge, 2017), 3. Also of note here is the curatorial sensibility of Daniela

Bleichmar's *Visible Empire: Botanical Expeditions and Visual Culture in the Hispanic Enlightenment* (Chicago: University of Chicago Press, 2012), which shows the importance of focusing on objects in the production and circulation of both knowledge and nature.

Chapter 1

1. Nathaniel Bagshaw Ward, *On the Growth of Plants in Closely Glazed Cases* (London: John Van Voorst, 1842), 26. On Ward living at 7 Wellclose Square, see Gerald Turner, *God Bless the Microscope! A History of the Royal Microscopical Society over 150 Years* (London: Royal Microscopical Society, 1989).

2. Nathaniel Bagshaw Ward, "Letter Addressed to R. H. Solly, Respecting His Method of Conveying Ferns and Mosses from Foreign Countries, and of Growing Them with Success in the Air of London," dated 9 December 1833, *Transactions of the Society for the Encouragement of Arts, Manufactures, and Commerce* 50 (1833): 226.

3. Description of Ward's house and quote from J. C. Loudon, "Growing Ferns and Other Plants in Glass Cases," *Gardener's Magazine*, April 1834, 162–63.

4. Ward, *On the Growth of Plants* (1842), 42.

5. After repeated visits by James McNab, of the Caledonian Horticultural Society, to Ward's house to see his cases, these specifications were given him by Ward, see Daniel Ellis, "Plant Case for Growing Plants without Water," *Gardener's Magazine*, September 1839, 481.

6. On early plant physiology, see Stephen Hales, *Vegetable Staticks* (London: W. and J. Innys, 1727); Joseph Priestley, "Observations on Different Kinds of Air," *Philosophical Transactions* 62 (1772), 198; Nicolas-Théodore de Saussure, *Recherches chimiques sur la végétation* (Paris: Nyon, 1804); Edward Turner and Robert Christison, "On the Effects of the Poisonous Gases on Vegetables," *Edinburgh Medical and Surgical Journal* 28 (1827): 356–63; and James Partington, *A Short History of Chemistry* (New York: Dover, 1989), 116.

7. Ward's early letters to societies include Ward to R. H. Solly, 9 December 1833, *Transactions of the Society, Instituted at London, for the Encouragement of Arts, Manufactures, and Commerce* 50, no. 1 (1833–34): 225–27; Ward to J. C. Loudon, 6 March 1834, *Gardener's Magazine* 10 (1834): 207–8; Ward to R. H. Solly, 29 December 1835, in *Transactions of the Society, Instituted at London, for the Encouragement of Arts, Manufactures, and Commerce* 50, no. 2 (1834–35): 190–91; and Ward to W. J. Hooker, 13 January 1836, in *Companion to the Botanical Magazine* 1 (1835): 317–20. On Faraday, see Michael Faraday to Ward, 4 November 1851, in N. B. Ward, *On the Growth of Plants in Closely Glazed Cases* (London: John Van Voorst, 1852), 133–34.

8. Ellis, "Plant Case," 505.

9. J. C. Loudon, "Calls at the London Nurseries, and Other Suburban Gardens," *Gardener's Magazine and Register of Rural and Domestic Improvement* 9 (1833): 467–68; A. R. P. Hayden, "Loddiges, George (1786–1846)," *Oxford Dictionary of National Biography*, rev. ed. (Oxford: Oxford University Press, 2004); and David Solman, *Loddiges of Hackney: The Largest Hothouse in the World* (London: Hackney Society, 1995).

10. Joseph Sabine, "Account of a Method of Conveying Water to Plants, in Houses, Invented by Mr. George Loddiges of Hackney," *Transactions of the Horticultural Society of London* 4 (1820): 14–16; and Solman, *Loddiges of Hackney,* 35–37.

11. Ward, *On the Growth of Plants* (1842), vi.

12. Ward to David Don, 4 June 1833, in Ward, *On the Growth of Plants* (1842), 75 76.

13. Livingstone, "Observations on the Difficulties."

14. For details on the *Persian,* see "Advertisement," *Public Ledger and Daily Advertiser,* 29 March 1833, 1; "Trade and Shipping," *Hobart Town Courier* 15 November 1833, 2; "Shipping Intelligence," *Sydney Herald,* 2 January 1834, 2; and "Shipping News," *The Australian,* 3 January 1834, 2.

15. Ward to Solly, 9 December 1833, 226–27.

16. Charles Mallard to Ward, 23 November 1833, DC, vol. 8, 153, RBGK. The *Persian* arrived in Hobart on 3 November 1833 and departed on 22 December 1833; see "Trade and Shipping," *Hobart Town Courier,* 15 November 1833, 2; and "Shipping Intelligence," *Sydney Herald,* 2 January 1834, 2.

17. The goods carried on the *Persian* are listed in "Imports," *Sydney Herald,* 6 January 1834, 2.

18. Walter Froggatt, "The Curators and Botanists of the Botanic Gardens, Sydney," *Journal and Proceedings of the Royal Australian Historical Society* 18 (1932): 101–33; and Charles Mallard to Ward, 18 January 1834, DC, vol. 8, 151, RBGK.

19. The ship sailed on 20 May 1834; see "Departures," *Sydney Herald,* 22 May 1834, 2. On details of the travel, see Ward to W. J. Hooker, 13 January 1836; and Ward, *On the Growth of Plants* (1842), 47.

20. Ward, *On the Growth of Plants* (1842), 47.

21. On Traill in Egypt, see Alix Wilkinson, "James Traill and William McCulloch: Two Nineteenth-Century Horticultural Society Gardeners in Egypt," *Garden History* 39 (2011): 83–98.

22. Plant list contained in Ward, *On the Growth of Plants* (1842), 80.

23. On William Light on the Nile, see David F. Elder, "Light, William (1786–1839)," in *Australian Dictionary of Biography* (London: Cambridge University Press, 1967), accessed 7 July 2017, http://adb.anu.edu.au/biography/light-william-2359/text3089. On Higgins's comments see Ward, *On the Growth of Plants* (1842), 79; see also *British Parliamentary Papers* 8 (1841).

24. James Traill to Ward, 30 April 1835, in Ward, *On the Growth of Plants* (1842), 79. On coffee, see Ward, *On the Growth of Plants* (1842), 48. On the success of plants in Egypt, see James Traill to John Bowring, 9 February 1838, in *Gardener's Magazine* 16 (1840), 652.

Chapter 2

1. Benjamin Franklin to Barbeu Duborg, 1773. Reproduced as Benjamin Franklin, "Restoration of Life by Sun Rays," in *Reading the Roots: American Nature Writing before Walden,* ed. Michael P. Branch (Athens: University of Georgia Press, 2004), 154–55.

2. Marie-Noelle Bourguet and Otto H. Sibum, introduction, in *Instruments, Travel and Science: Itineraries of Precision from the Seventeenth to the Twentieth Century* (London: Routledge, 2002), 1–19.

3. Boyle's comments were first published in 1666 and later published in book form as Robert Boyle, *General Heads for the Natural History of a Country, Great or Small: Drawn Out for the Uses of Travellers and Navigators* (London: John Taylor, 1692).

4. "I Send you . . . ," quoted in "An Accurate Description of the Cacao Tree," *Philosophical Transactions* 8, no. 93 (1673): 6007. On introductions by the end of the seventeenth century, see Desmond, "Technical Problems," 75.

5. On Evelyn, see Kenneth Lemmon, *The Golden Age of Plant Hunters* (New York: A. S. Barnes, 1968), 7.

6. James Woodward, *Brief Instructions for Making Observations in All Parts of the World; as also for Collecting, Preserving and Sending Over Natural Things* (London: Richard Wilkin, 1696), 13. On Woodward, see also Desmond, "Technical Problems," 75.

7. Reproduced in James Petiver, *Jacobi Petiveri Opera, Historiam Naturalem Spectantia*, vol. 1 (London: John Millan, 1767). See also James Delbourgo, "Listing People," *Isis* 103, no. 4 (2012): 735–42.

8. John Collinson to John Bartram, 20 January 1735. In Alan W. Armstrong, ed. *"Forget not Mee & My Garden . . .": Selected Letters 1725–1768 of Peter Collinson, F.R.S.* (Philadelphia: American Philosophical Society, 2002), 22–23.

9. John Collinson to John Bartram, 24 January 1734, in Armstrong, *"Forget not Mee"*, 11–13.

10. Desmond, "Technical Problems," 79.

11. On Bartram's boxes, see *Gentlemen's Magazine*, 24 February 1754, 65. See also Andrea Wulf, *The Brother Gardeners: Botany, Empire and the Birth of an Obsession* (London: Windmill, 2009), 137–38; and Kathleen Clark, "What the Nurserymen Did for Us: The Roles and Influence of the Nursery Trade on the Landscapes and Gardens of the Eighteenth Century," *Garden History* 40, no. 1 (2012): 17–33, at 26.

12. John Bartram to Peter Collinson, 27 April 1755, in *Memorials of John Bartram and Humphrey Marshall*, ed. William Darlington (Philadelphia: Lindsay & Blakiston, 1849), 200.

13. Henri-Louis Duhamel du Monceau, quoted in Christopher M. Parsons and Kathleen S. Murphy, "Ecosystems under Sail: Specimen Transport in the Eighteenth-Century French and British Atlantics," *Early American Studies* 10, no. 3 (2012): 503–39, at 504.

14. Quoted in ibid., 508.

15. Richard Grove, *Green Imperialism: Colonial Expansion, Tropical Island Edens and the Origins of Environmentalism, 1600–1860* (Cambridge: Cambridge University Press, 1995), 42–43.

16. John Fothergill, *Directions for Taking Up Plants and Shrubs, and Conveying Them by Sea* (ca. 1770), quoted in Desmond, "Technical Problems," 81. The same plant boxes also appear in John Coakley Lettsom, *Hortus Uptonensis; or, a Catalogue of Stove*

NOTES TO PAGES 36–43   233

and Green-House Plants in Dr. Fothergill's Garden at Upton, at the Time of his Decease (London, 1783).

17. Christopher J. Humphries and Robert Huxley, "Carl Linnaeus: The Man Who Brought Order to Nature," in *The Great Naturalists*, ed. Robert Huxley (London: Thames & Hudson, 2007), 132–39.

18. Roy A. Rauschenberg, "John Ellis, F.R.S.: Eighteenth Century Naturalist and Royal Agent to West Florida," *Notes and Records of the Royal Society of London* 32, no. 2 (1978): 149–64.

19. Some of these include "An Account of Some Experiments Relating to the Preservation of Seeds," *Philosophical Transactions* 51 (1759–1760): 206–15; and "A Letter from John Ellis, Esquire, F.R.S. to the President, on the Success of His Experiments for Preserving Acorns for a Whole Year without Planting Them," *Philosophical Transactions* 58 (1769): 75–79.

20. John Ellis, *Directions for Bringing Over Seeds and Plants from the East-Indies and Other Distant Countries in a State of Vegetation* (Davis: London, 1770), 1.

21. Ibid., 7.

22. Ibid., 10.

23. Ibid., 21.

24. John Ellis, *A Description of the Mangostan and the Bread-Fruit* (London: John Ellis, 1775), 20.

25. "The most delicious" and "the most useful," from the extended title of ibid.

26. Ibid., 19.

27. Parsons and Murphy, "Ecosystems under Sail," 538.

28. André Thouin, *Cours de culture et de naturalisation des végétaux: Atlas* (Paris: Huzard, 1827).

29. Ray Desmond, *The History of the Royal Botanic Gardens Kew*, 2nd ed. (Kew: Royal Botanic Gardens, Kew, 2007), 107.

30. Richard Sorrenson, "The Ship as a Scientific Instrument in the Eighteenth Century," *Osiris* 11 (1996): 221–36.

31. Rigby, "Politics and Pragmatics of Seaborne Plant Transportation," 81.

32. David Mackay, *In the Wake of Cook: Exploration, Science and Empire* (Wellington: Victoria University Press, 1985), 123.

33. Lord Sydney to Joseph Banks, 15 August 1787; this letter appears in Allan Frost, *Sir Joseph Banks and the Transfer of Plants to and from the South Pacific, 1786–1798* (Melbourne: Colony Press, 1993), 47. For an alternative and illustrative story of the breadfruit transplant as seen from Jamaica, see Elizabeth DeLoughrey, "Globalizing the Routes of Breadfruit and Other Bounties," *Journal of Colonialism and Colonial History*, 8, no. 3 (2007), https://muse.jhu.edu/article/230160.

34. Much of this section relies upon Frost, *Sir Joseph Banks*, 37–38. On the number of trees, see Rigby, "Politics and Pragmatics of Seaborne Plant Transportation," 95.

35. Banks to Grenville, 7 June 1789, in Frost, *Sir Joseph Banks*, 22; Banks to Riou, 5 June 1789, in Frost, *Sir Joseph Banks*, 21.

36. Joseph Banks, "Instructions to the Gardeners on the *Guardian*," July 1789, London, in Frost, *Sir Joseph Banks*, 25–26.

37. On the plant cabin overboard, see Desmond, *Kew*, 119.

38. "Sketch Porpoise Sloop's Quarterdeck, showing the manner in which the Garden Cabbin was fitted with Boxes, agreeable to Sir Joseph Banks's desire for the reception of Plants to be sent to Port Jackson," n.d., ms. series 19.46, State Library of NSW, http://www2.sl.nsw.gov.au/banks/series_19/19_46.cfm.

39. H. B. Carter, *Sir Joseph Banks, 1743–1820* (London: British Museum, 1988), 558.

40. Livingstone, "Observations on the Difficulties."

41. John Lindley, "Instructions for Packing Living Plants in Foreign Countries," *Transactions of the Horticultural Society of London* 5 (1824): 192–200; and William Stearn, *John Lindley, 1799–1865: Gardener, Botanist and Pioneer Orchidologist* (Woodbridge: Antique Collectors' Club, 1999).

42. On Reeves's boxes see "Obituary for John Reeves," *Gardeners' Chronicle*, 29 March 1856, 212. Farquhar's box is more fully described in John Lindley and J. C. Loudon, "Implements of Gardening," in *An Encyclopedia of Gardening Comprising the Theory and Practice of Horticulture, Floriculture, Arboriculture, and Landscape-Gardening* (London: Longman, Rees, Orme, Brown, Green & Longman, 1835), 585–86.

43. Joseph Dalton Hooker, "Obituary: N. B. Ward," *Gardeners' Chronicle*, 20 June 1868, 655–56.

Chapter 3

1. Quoted in W. J. Hooker to Ward, 4 April 1851, in Ward, *On the Growth of Plants* (1852), 131–32; and George Loddiges to Ward, 18 February 1842, in ibid., 123–25.

2. "The Late Mr. George Loddiges," *Journal of the Horticultural Society of London* 1 (1848): 224.

3. More than a nursery owner, Loddiges was also the leading authority on hummingbirds. When consignments of plants were sent to the nursery, Loddiges requested that collectors place hummingbirds in specially constructed drawers at the bottom of plant transporting cases. See "Miscellaneous bits of paper relating to collection," Box 2, Manuscript Collection of George Loddiges, Natural History Museum, London.

4. Kate Colquhoun, *"The Busiest Man in England": A Life of Joseph Paxton, Gardener, Architect and Victorian Visionary* (Boston: Godine, 2006); Tatiana Holway, *The Flower of Empire: An Amazonian Water Lily, the Quest to Make It Bloom, and the World It Created* (New York: Oxford University Press, 2013), 110–11.

5. Nathaniel Wallich, "The Discovery of the Tea Shrub in India," *Gardener's Magazine* 15 (1835): 428–30.

6. Quoted in Lemmon, *Golden Age*, 187.

7. Quoted in Colquhoun, *"The Busiest Man,"* 69 and 70.

8. Details of Maulee and Baugh, in Nathaniel Wallich to H. T. Prinsep, 9 August 1837, India Office Records (IOR), P/13/24, nos. 46–48, British Library, Archives, London, accessed 20 September 2017, http://www.bl.uk/manuscripts/Viewer.aspx?ref=ior!p!13!24_9_Aug_1837_nos_46-48_f001r#.

9. Wallich, quoted in Lemmon, *Golden Age*, 211.

10. "The plants are . . . ," quoted in Kenneth Lemmon, *The Covered Garden* (London: Museum Press, 1962), 205–6. See also "John Gibson," *Gardeners' Chronicle*, 29 June 1872, 865; and Alice Coats, *The Plant Hunters: Being a History of the Horticultural Pioneers, Their Quests and Their Discoveries from the Renaissance to the Twentieth Century* (New York: McGraw-Hill, 1969), 154–55.

11. W. J. Hooker, *Directions for Collecting and Preserving Plants in Foreign Countries* (London, 1828), 115. Hooker would not include the Wardian case until 1842; see W. J. Hooker, *A Few Plain Instructions for Collecting and Transporting Plants in Foreign Countries* (London: The Admiralty, 1848).

12. Harvey to W. J. Hooker, 12 December 1834, in William Henry Harvey, *Memoir of W. H. Harvey* (London: Bell & Dalby, 1869), 48–49.

13. William J. Hooker, "Wardia: A New Genus of Mosses, Discovered in Southern Africa," *Companion to the Botanical Magazine* 2 (1836): 183–85.

14. "Letter from N. B. Ward Esq. to Dr. Hooker, on the Subject of His Improved Method of Transporting Living Plants," *Companion to the Botanical Magazine* 1 (1836): 317–20.

15. "I believe . . . ," in Ward to W. J. Hooker, undated (postmarked 19 April 1836), fol. 150, DC, RBGK. See also Ward to W. J. Hooker, 24 May 1836, fol. 154, in DC, RBGK; Ward to W. J. Hooker, 4 September 1836, fol. 324, in DC, RBGK.

16. "It may be . . . ," in Ward to W. J. Hooker, undated (postmarked 19 April 1836), ibid.

17. Gardner to W. J. Hooker, 5 May 1840, and Gardner to W. J. Hooker, 18 December 1840, in *Journal of Botany* 3 (1841): 134–37 and 201–3.

18. G. S. Boulger, "Gardner, George (1812–1849)," in *Oxford Dictionary of National Biography* (Oxford: Oxford University Press, 2004), accessed 26 November 2019, https://doi.org/10.1093/ref:odnb/10373. See also George Gardner, "Medicinal and Economic Plants, Fruits &c of the North of Brazil," Manuscripts of George Gardner, RBGK.

19. "Petersburg Botanic Garden," *American Journal of Science and Arts* 20, no. 1 (1831): 175–76; and Heldur Sander, Toivo Meikar, and Anita Magowska, "The Learned Gardeners of the Botanical Gardens of the University of Tartu and their Activities (1803–1918)," *Acta Baltica Historiae et Philosophiae Scientiarum* 2, no. 1 (2014): 53–110.

20. On the boxes according to Ward's plan, see Gardner to W. J. Hooker, 23 March 1841, in *Journal of Botany* 4 (1842): 199–201. On *Prepusa* and collecting orchids, see Gardner to W. J. Hooker, 5 May 1841, in *Journal of Botany* 4 (1842): 201–2. See also George Gardner, *Travels in the Interior of Brazil Principally through the Northern Provinces and the Gold and Diamond Districts during the Years 1836–1841* (London: Reeve, Benham & Reeve, 1849).

21. Gardner to W. J. Hooker, 22 May 1841, in *Journal of Botany* 4 (1842): 202–3. See also Daniel Domingues da Silva, "The Atlantic Slave Trade to Maranhão, 1680–1846: Volume, Routes and organisation," *Slavery and Abolition* 20, no. 4 (2008): 477–501.

22. Gardner to W. J. Hooker, 6 July 1841, in *Journal of Botany* 4 (1842): 204–5.

23. W. J. Hooker, editorial comment in "Contributions towards a Flora of Brazil Being an Enumeration of a Series of Collections of Plants, Made in Various Parts of Brazil, from 1836 to 1841," *London Journal of Botany* 1 (1842) 158–93, at 165.

24. Ward to W. J. Hooker, 16 May 1837, fol. 325, in DC, RBGK.

25. This and the next paragraph: James Yates, "Report of the Committee for Making Experiments on the Growth of Plants under Glass, and without Any Free Communication with the Outward Air; on the plan of Mr N. B. Ward of London," in *Report of the British Association for the Advancement of Science*, vol. 6 (London: Richard & John E. Taylor, 1838), 501–8. On details of the events at the meeting, see also "Meeting of the British Association," *Manchester Courier*, 16 September 1837, 4; and "Meeting of the British Association at Liverpool," *Wolverhampton Chronicle and Staffordshire Advertiser*, 20 September 1837, 3. See also Charles Withers, Rebekah Higgitt, and Diarmid Finnegan, "Historical Geographies of Provincial Science: Themes in the Setting and Reception of the British Association for the Advancement of Science in Britain and Ireland, 1831–c. 1939," *British Journal for the History of Science* 41, no. 3 (2008): 385–415.

26. N. B. Ward, "Mr. Ward's Report," in Yates, "Report of the Committee," 503.

27. Ellis, "Plant Case": description of the box, 481–86; "zeal . . . ," 504; "acquired . . . ," 505.

28. Allan Maconochie, "On the Use of Glass Cases for Rearing Plants Similar to Those Recommended by N. B. Ward," *Annual Report and Proceedings of the Botanical Society of Edinburgh* 3 (1840): 96–97; Allan Maconochie, "Notice Regarding the Growth of Plants in Close Glazed Cases," *Proceedings of the Royal Society of Edinburgh* 1 (1845): 299. On the Maconochie incident, see also Allen, *The Victorian Fern Craze*.

29. "Robberies, Accidents, &c." *Bell's New Weekly Messenger*, 10 November 1839, 7.

30. Ward to W. J. Hooker, 21 April 1840, fol. 216, in DC, RBGK.

31. Ibid. Note that Ward completed the book well before it was published in 1842.

32. Ward, *On the Growth of Plants* (1842).

33. John Lindley, "Botanical Garden (Kew): Copy of the Report Made to the Committee Appointed by the Lords of the Treasury in January 1838 to Inquire into the Management, &c. of the Royal Gardens at Kew," in *House of Commons Parliamentary Papers*, vol. 29, Paper No. 292 (1840).

34. For details on plant dispersals and quotations, see Lindley, ibid., 3, 4, and 5.

35. Aiton to Lindley, 22 February 1838, in ibid., 3–4. On the first case sent from Kew, see Royal Botanic Gardens, Kew, Kew Plant Record Books, Outwards Goods, 1836–1847, p. 55, RBGK.

36. The years 1838–41 have been described by others: Desmond, *Kew*; Guy Meynell, "Kew and the Royal Gardens Committee of 1838," *Archives of Natural History* 10, no. 3 (1982): 469–77; and William Stearn, "The Self-Taught Botanists Who Saved the Kew Botanic Garden," *Taxon* 14, no. 9 (1965): 293–98.

37. Case sent to Ward, see Royal Botanic Gardens, Kew, Kew Plant Record Books, Outwards Goods, 1836–1847, p. 98, RBGK. Cases sent by Symonds, including Joseph Hooker's note, Royal Botanic Gardens, Kew, Kew Plant Record Books, Inwards Goods,

1828–1847, pp. 85–86, RBGK. See also William Symonds to W. J. Hooker, 17 April and 26 April 1842, Folio 161, DC, RBGK.

Chapter 4

1. The Hackney tour took place on 2 February 1839. Much of this discussion taken from Asa Gray, "First Journey in Europe, 1838–1839," in *Letters of Asa Gray*, vol. 1, ed. Jane Loring Gray (Boston: Houghton, Mifflin, 1894), 85–271; trip to Loddiges recounted, 126–27.

2. "Attracted much . . . ," in Gray, *Letters of Asa Gray*, 23. Further quotes: "plant case . . . ," 126; "one of . . . ," 126; "his house . . . ," 126; in Gray, *Letters of Asa Gray*. On Harvard appointment, see A. Hunter Dupree, *Asa Gray, 1810–1888* (Cambridge, MA: Belknap Press, 1959).

3. J. D. Hooker to Ward, 26 November 1842, in Joseph Hooker: Correspondence 1839–1845 from Antarctic Expedition (hereafter cited as Hooker Correspondence), Archives of the RBGK.

4. Mallard's original letters are held in RBGK. See Charles Mallard to Ward, 18 January 1834, DC, fol. 151, RBGK.

5. Loddiges to Ward, 18 February 1842, in Ward, *On the Growth of Plants* (1842), 86–87.

6. Nathaniel Wallich. "Upon the Preparation and Management of Plants during a Voyage from India," *Transactions of the Horticultural Society of London* 1, 2nd ser. (1832): 140–43; see also Ray Desmond, *The European Discovery of Indian Flora* (Oxford: Oxford University Press, 1992), 319. On the more theoretical ideas behind these "spaces for natural history," see Dorinda Outram, "New Spaces in Natural History," in *Cultures of Natural History*, ed. N. Jardine, J. A. Secord, and E. C. Spray (Cambridge: Cambridge University Press, 1996), 447–59; Sorrenson, "The Ship as a Scientific Instrument"; and Parsons and Murphy, "Ecosystems under Sail."

7. For a short discussion of the three expeditions see Stephen J. Pyne, *The Ice* (London: Phoenix, 2004), 74–81. Among the many excellent studies of the individual expeditions, see Edward Duyker, *Dumont d'Urville: Explorer and Polymath* (Honolulu: University of Hawai'i Press, 2014), 314–489; William Stanton, *The Great United States Exploring Expedition of 1838–1842* (Berkeley: University of California Press, 1975); Herman J. Viola and Carolyn Margolis, eds., *Magnificent Voyagers: The U.S. Exploring Expedition, 1838–1842* (Washington, DC: Smithsonian Press, 1985); and Maurice J. Ross, *Ross in the Antarctic: The Voyages of James Clark Ross in Her Majesty's Ships Erebus and Terror, 1839–1843* (Whitby: Caedmon of Whitby, 1982).

8. First French case described in de Mirbel, Cordier, de Blainville, de Freycinet, and Savary, "Rapport de la Commission chargée, sur l'invitation de M. le Ministre de la Marine, de rédiger des instructions pour les observations scientifiques à faire pendant le voyage des corvettes de l'État l'Astrolabe et la Zélée, sous le commandement de M. le capitaine Dumont d'Urville," in *Comptes rendus hebdomadaires des séances de l'Académie des Sciences*, 5 (1837): 133–55; botanical instructions by Mirbel, 134–42. Translated in

part in English as Charles de Mirbel, "Instructions for Preserving Plants," *Nautical Magazine and Naval Chronicle* 3 (1838): 164–70.

9. The following discussion is taken from Mirbel, "Instructions," 164–68.

10. *Le moniteur universel*, quoted in Duyker, *Dumont d'Urville*, 87.

11. *Annales maritimes et colonials*, quoted in ibid., 479.

12. Jacques Bernard Hombron, Joseph Decaisne, Dumont d'Urville, Charles Jacquinot, and Jean Montagne, *Botanique*, 3 vols. (Paris: Gide et Cie, 1845–53); and F. Bruce Sampson, *Early New Zealand Botanical Art* (Auckland: Reed Methuen, 1985), 53–57. *Acaena* is now known as *Acaena anserinifolia*.

13. J. D. Hooker to Hector, 14 December 1884, in *My Dear Hector: Letters from Joseph Dalton Hooker to James Hector, 1862–1893*, ed. Juliet Hobbs (Wellington: Museum of New Zealand, 1999), 181–82.

14. Royal Society, *Report of the President and Council of the Royal Society on the Instructions to be Prepared for the Scientific Expedition to the Antarctic Regions* (London: Taylor, 1839), 36.

15. "I felt . . . ," in Ward to J. D. Hooker, 26 September 1839, in Joseph Hooker, Correspondence Received 1839–1845, fol. 256, RBGK. On Hooker's equipment, see Joseph Hooker, *Life and Letters of Sir Joseph Dalton Hooker* (London: John Murray, 1918), 47.

16. J. D. Hooker (JDH) to W. J. Hooker (WJH), 17 March 1840, in Hooker Correspondence, fols. 26–27, RBGK.

17. "Long consult . . . ," "the late . . . ," in JDH to WJH, 7 September 1840; on surviving plants, see JDH to WJH, 9 November 1840, and JDH to WJH, 16 August 1840; all in Hooker Correspondence, fols. 37, 45, and 31.

18. JDH to WJH, 9 November 1840, in Hooker Correspondence, fol. 45.

19. Jim Endersby, *Imperial Nature: Joseph Hooker and the Practices of Victorian Science* (Chicago: University of Chicago Press, 2008); Joseph Hooker, *Flora Tasmaniae* (London: Lovell Reeve, 1860). On algae, see Ward to WJH, 31 December 1839, DC, fol. 156, RBGK; Ward to JDH, 26 September 1839, in Hooker, Correspondence Received, fol. 256; JDH to Ward, 26 November 1842, in Hooker Correspondence, fols. 138–39.

20. JDH to WJH, 23 November 1841, in Hooker Correspondence, fol. 80. On not reaching Kew, see Record of Plants, Inwards, 1828–1847, RBGK, 85. The first recorded plants from JDH were from the Falkland Islands; see below.

21. Hooker's remark about cattle is quoted in James Ross, *A Voyage of Discovery and Research in the Southern and Antarctic Regions, during the Years 1839–1843*, vol. 2 (London: John Murray, 1847), 272. On cases received at Kew on 12 March 1843, see Record of Plants, Inwards, 1828–1847, RBGK, 117. "Flourished . . . ," in Joseph Hooker, *The Botany of the Antarctic Voyage*, vol. 1 (London: Reeve Brothers, 1844), 375. On grasses and empire, see Eric Pawson and Tom Brooking, *Seeds of Empire: The Environmental Transformation of New Zealand* (London: I. B. Taurus, 2011).

22. JDH to WJH, 20 April 1843, in Hooker Correspondence, fols. 196–200.

23. Cases received at Kew on 16 January 1844, see Record of Plants, Inwards, 1828–1847, 136, RBGK.

24. Stanton, *Great United States*, 49; Richard W. Blumenthal, *Charles Wilkes and the*

Exploration of Inland Washington Waters: Journals from the Expedition of 1841 (Jefferson, NC: McFarland, 2009); Doris E. Borthwick, "Outfitting the United States Exploring Expedition: Lieutenant Charles Wilkes' European Assignment, August–November, 1836," *Proceedings of the American Philosophical Society* 109 (1965): 159–72. There was some controversy regarding Wilkes sighting land; see Ross, *Ross in the Antarctic*, 118–32.

25. Charles Wilkes, *Autobiography of Rear Admiral Charles Wilkes, 1798–1877* (Washington, DC: Department of the Navy, 1978), 528.

26. Tyler, *Wilkes Expedition*; Stanton, *Great United States*; J. C. Loudon, "Mr Brackenridge," *Gardener's Magazine*, 10 (1834): 162. Donald Culross Peattie, "William Dunlop Brackenridge," *Dictionary of American Biography*, vol. 2 (New York: Scribner, 1928), 545.

27. William Brackenridge, "Journal of William Dunlop Brackenridge: October 1–28, 1841," *California Historical Society Quarterly* 24 (1945): 326–36, at 326–27.

28. Harley Harris Bartlett, "The Reports of the Wilkes Expedition, and the Work of the Specialists in Science," *Proceedings of the American Philosophical Society* 82, no. 5 (1940): 676–79; John Torrey, *On the* Darlingtonia californica, *a new pitcher-plant from Northern California* (Washington, DC: Smithsonian Institution, 1853). Mariana Bornholdt, "Botanizing Western Oregon in 1841—The Wilkes Inland Expedition," *Kalmiopsis* 12 (2005): 16–24. On the state of the national collection before the Wilkes expedition, see J. J. Abert, A. O. Dayton, Francis Markof, "Note A: January 1, 1842," in *A Memorial of George Brown Goode Together with a Selection of His Papers on Museums and the History of Science in America* (Washington, DC: Government Printing Office, 1901), 157–61.

29. Antony Adler, "From the Pacific to the Patent Office: The US Exploring Expedition and the Origins of America's First National Museum," *Journal of the History of Collections* 23 (2011): 49–73.

30. Charles Wilkes, *Autobiography*, 530.

31. Charles Wilkes, *A Brief Account of the Discoveries and Results of the United States Exploring Expedition* (New Haven: Hamlen, 1843). William D. Brackenridge, "Botanical Department," report presented to the National Institute, November 1842, republished in *A Memorial of George Brown Goode Together with a Selection of His Papers on Museums and the History of Science in America* (Washington, DC: Government Printing Office, 1901), 164. Bartlett, "Reports of the Wilkes Expedition," 676.

32. Quoted in Brackenridge, "Botanical Department," 164.

33. The Editor [Charles Mason Hovey], "Experimental Garden of the National Institute (Containing the Plants Collected on the Wilkes Expedition)," *Magazine of Horticulture and Botany* 10 (1844): 81–82; and "Notes of a Visit to Several Gardens in the Vicinity of Washington, Baltimore, Philadelphia, and New York, in October 1845," *Magazine of Horticulture and Botany* 12 (1846): 241–42. See also "The Patent Office Greenhouse," *Daily Union*, 26 June 1845, 194.

34. It is important to note here that, although the grounds at the foot of the Capitol had earlier roots as the Botanic Garden of the Columbia Institute, there were ten times as many plants in the Patent Office greenhouse as had existed in the earlier garden, and without the transfer of Brackenridge's plants, the Columbia Institute Garden would

have fallen into ruin; see Anne-Catherine Fallen, *A Botanic Garden for the Nation: The United States Botanic Garden* (Washington, DC: Government Printing Office, 2007). Fallen also details the plants that are still in the gardens that were descended from the Wilkes expedition (next paragraph). The USBG was in continuous operation at the foot of the Capitol until 1933, when it was moved to its new location on Independence Avenue.

35. Brackenridge, "Botanical Department," 164–65.

36. Endersby, *Imperial Nature*, 59; and Desmond, *Kew*, 195–96.

37. William Brackenridge to Asa Gray, 3 May 1850, in William Brackenridge Letters, 1850–1856 (hereafter cited as HL Brack 1), Gray Herbarium Archives, Harvard University.

38. Rigby, "Politics and Pragmatics of Seaborne Plant Transportation," 97. See also F. E. *A Popular Account of Their Construction. Development, Management and Appliances* (New York: Scribner, 1891).

Chapter 5

1. Keogh, "The Wardian Case: Environmental Histories." On the banana see Ward to J. J. Bennett, 1 November 1842, in *Proceedings of the Linnean Society of London* 1 (1848): 157. On gutta percha see "Our Book Shelf," *Nature* 61 (1909): 538. On the mango see "The Mango of Queensland," *Brisbane Courier*, 12 March 1870; "Queensland Acclimatisation Society," *Queenslander*, 3 February 1877; and Jodi Frawley, "Making Mangoes Move," *Transforming Cultures* 3 (2008): 165–84.

2. Historians often like to label it "the tea war"; see Markman Ellis, Richard Coulton, and Matthew Mauger, *Empire of Tea: The Asian Leaf That Conquered the World* (London: Reaktion, 2015), 213–19. For a much fuller account of the Opium War, see Mao Haijian, *The Qing Empire and the Opium War: The Collapse of the Heavenly Dynasty* (Cambridge: Cambridge University Press, 2016).

3. Robert Fortune wrote four popular accounts of his travels: *Three Years' Wanderings in the Northern Provinces of China* (London: John Murray, 1847); *A Journey to the Tea Countries of China* (London: John Murray, 1852); *A Residence among the Chinese: Inland, on the Coast, and at Sea* (London: John Murray, 1857); and *Yedo and Peking: A Narrative of a Journey to the Capitals of Japan and China* (London: John Murray, 1863). Modern literature on Fortune is substantial; of note is the recent biography by Alistair Watt, *Robert Fortune: A Plant Hunter in the Orient* (Kew: Royal Botanic Gardens, Kew, 2017).

4. Robert Fortune, "Ward's Plant Cases," *Gardeners' Chronicle*, 27 June 1868, 608.

5. During the committee's meetings much discussion was given to Wardian cases. See Minutes of the Chinese Committee, Folder 1, Papers of Robert Fortune, Royal Horticultural Society, London (hereafter cited as Fortune Papers).

6. Instructions to Mr Robert Fortune Proceeding to China in the Service of the Horticultural Society of London, Folder 1, Fortune Papers. The results of the experiment were reported in Robert Fortune, "Experience in the Transmission of Living Plants To and From Distant Countries by Sea," *Journal of the Horticultural Society of London* 2 (1847): 115–21.

7. For a thorough description of Fortune's introductions, see Watt, *Robert Fortune,* 29–89.

8. Fortune, *Three Years' Wanderings,* 411.

9. A number of Fortune's claimed discoveries were well known to others who were stationed in China; see Samuel Ball, *An Account of the Cultivation and Manufacture of Tea in China: Derived from Personal Observation during an Official Residence in that Country from 1804 to 1826* (London: Longman, Brown, Green & Longman, 1848).

10. Described in Fortune, *Three Years' Wanderings,* 186–206; and Watt, *Robert Fortune,* 253–71.

11. Fortune describes Ward as "my old friend," in Fortune, *Yedo and Peking,* 147.

12. Wallich, "Discovery of the Tea Shrub," 429. Tea was discovered in Assam in 1823 by Robert Bruce; see Ellis, Coulton, and Mauger, *Empire of Tea,* 209–10.

13. Described in Fortune, *Tea Countries,* 355–57.

14. Quoted in ibid., 356.

15. Ward describes the method in the first edition of his book *On the Growth of Plants* (1842), 51.

16. The contents of the cases are described in Robert Fortune, List of Tea Plants, Seeds and Implements Sent, in Saharanpur Botanic Garden, including papers re tea cultivation, ms. IOR/F/4/2498/141673, p. 80 [fol. 39v], Digitised Manuscripts, British Library, accessed 16 July 2018, http://www.bl.uk/manuscripts/FullDisplay.aspx?ref=IOR/F/4/2498/141673 (hereafter cited as Saharanpur Botanic Garden ms.).

17. Fortune, *Tea Countries,* 357.

18. Contract with Chinese Workers, 22 February 1851, Saharanpur Botanic Garden ms., pp. 81–83 (fol. 40r–fol. 41r).

19. Robert Fortune, Implementation for the Manufacture of Tea, Saharanpur Botanic Garden ms., p. 79 (fol. 39r).

20. Fortune to Falconer, 30 December 1853, in Cases of Tea Seedlings, ms. IOR/P/14/39, Digitised Manuscripts, British Library, accessed 26 November 2019, http://www.bl.uk/manuscripts/FullDisplay.aspx?ref=IOR/P/14/39_16_Mar_1854_nos_101-107; Hugh Falconer to Under Secretary, 7 March 1854, in Cases of Tea Seedlings.

21. List of plants introduced into the Calcutta Botanic Garden from China by Robert Fortune, in Calcutta Botanic Garden, ms. IOR/P/14/68, Digitised Manuscripts, British Library, accessed 26 November 2019, http://www.bl.uk/manuscripts/Viewer.aspx?ref=ior!p!14!68_16_Oct_1856_nos_51-52_f001r.

22. Watt, *Robert Fortune,* 269–70.

23. Percival Griffiths, *The History of the Indian Tea Industry* (London: Weidenfeld & Nicholson, 1967).

24. Haripriya Rangan, "State Economic Policies and Changing Regional Landscapes in the Uttarakhand Himalaya, 1818–1947," in *Agrarian Environments: Resources, Representations, and Rule in India,* ed. Arun Agrawal and K. Sivaramakrishnan (Durham, NC: Duke University Press, 2000), 23–46.

25. Gadapani Sarma, "A Historical Background of Tea in Assam," *Echo* 1, no. 4 (2013): 123–31.

26. Quoted in William Gardener, "Robert Fortune and the Cultivation of Tea in the United States," *Arnoldia* 31, no. 1 (1971): 1–18, at 4.

27. The description of this journey is taken from ibid., 6–8.

28. D. J. B., "Preparation for a Government Propagating Garden at Washington," in *Report of the Commissioner of Patents for the Year 1858: Agriculture*, ed. J. Holt (Washington, DC: Steedman, 1859), 280–83, at 282.

29. This point is also made by the tea planter Roy Moxham in *Tea: A History of Obsession, Exploitation, and Empire* (New York: Carroll & Graf, 2003).

30. Mathew J. Crawford, "Between Bureaucrats and Bark Collectors: Spain's Royal Reserve of Quina and the Limits of European Botany in the Late Eighteenth-Century Spanish Atlantic World," in *Global Scientific Practice in an Age of Revolutions, 1750–1850*, ed. Patrick Manning and Daniel Rood (Pittsburgh: University of Pittsburgh Press, 2016), 21–37; and I. W. Sherman, "A Brief History of Malaria and Discovery of the Parasite's Life Cycle," in *Malaria: Parasite Biology, Pathogenesis and Protection*, ed. I. W. Sherman (Washington, DC: ASM, 1998), 3–10.

31. Clements Markham, *Travels in Peru and India: While Superintending the Collection of Chinchona Plants and Seeds in South America, and Their Introduction into India* (London: John Murray, 1862), 334–35.

32. Arjo Roersch van der Hoogte and Toine Pieters, "Science in the Service of Colonial Agro-Industrialism: The Case of Cinchona Cultivation in the Dutch and British East Indies, 1852–1900," *Studies in History and Philosophy of Biological and Biomedical Sciences* 47 (2014): 12–22. Kim Walker and Mark Nesbitt, *Just the Tonic: A History of Tonic Water* (Kew: Royal Botanic Gardens, Kew, 2019).

33. Crawford, "Between Bureaucrats and Bark Collectors," 24. The cinchona initiative has received wide attention; see, for example, Mark Honigsbaum, *The Fever Trail: The Hunt for the Cure for Malaria* (New York: Farrar, Straus & Giroux, 2002); Lucille Brockway, *Science and Colonial Expansion: The Role of the British Royal Botanic Gardens* (New York: Academic Press, 1974), 103–40; and Henry Hobhouse, *Seeds of Change: Five Plants That Transformed Mankind* (London: Sidgwick & Jackson, 1985). For a succinct account, see Richard Drayton, *Nature's Government: Science, Imperial Britain, and the "Improvement" of the World* (New Haven, CT: Yale University Press, 2000), 206–11.

34. "Chinakultur auf Java," *Botanische Zeitung*, 30 June 1865, 208–11; "Obituary: Hasskarl," *Chemist and Druggist*, 20 January 1894, 73–74; and Norman Taylor, *Cinchona in Java: The Story of Quinine* (New York: Greenberg, 1945), 38.

35. The Dutch venture is described more fully in Andrew Goss, *The Floracrats: State-Sponsored Science and the Failure of the Enlightenment in Indonesia* (Madison: University of Wisconsin Press, 2011), 33–58.

36. The collectors are described in Daniel R. Headrick, *Power over Peoples: Technology, Environments, and Western Imperialism, 1400 to the Present* (Princeton, NJ: Princeton University Press, 2010). 233.

37. Minna Markham Diary, entry for 9 October 1860; quoted in Lucy Veale, "An Historical Geography of the Nilgiri Cinchona Plantations, 1860–1900" (PhD diss., University of Nottingham, 2010), 151.

38. Minna Markham Diary, entry for 11 October 1860; quoted in ibid., 151.

39. "Mr Spruce's Report on the Expedition to Procure Seeds and Plants of the Cinchona succirubra, or Red Bark Tree, to the Under Secretary of State for India, 3rd January 1862," in British Parliamentary Papers, *Copy of Correspondence Relating to the Introduction of the Chinchona Plant into India, and to Proceedings Connected with Its Cultivation, from March 1852 to March 1863* (hereafter cited as BPP, *Introduction of Chinchona*) (London: House of Commons, 1863), 65–118, at 99; and Richard Spruce, *Notes of a Botanist on the Amazon and Andes* (London: Macmillan, 1908), 258–311. On collecting the orchid, see Royal Botanic Gardens, Kew, Kew Plant Record Books, Inwards 1859–1867, ms. RBGK, p. 92. On the number of plants propagated, see Thomas Anderson, "General Report on the Cultivation of the Species Cinchona in the Neilgherries," in BPP, *Introduction of Chinchona*, 207.

40. T. C. Owen, *The Cinchona Planter's Manual* (Colombo: A. M. & J. Ferguson, 1881), 93.

41. T. Anderson, "Report on the Experimental Cultivation of the Quiniferous Chinchonae in British Sikhim," in BPP, *Introduction of Chinchona*, 260.

42. Taylor, *Cinchona in Java*, 43.

43. Ibid.

44. Thomas Anderson to W. Grey, in BPP, *Introduction of Chinchona*, 189–90.

45. Mamani and Ledger's story is described in Gabrielle Grammacia, *The Life of Charles Ledger, 1818–1905: Alpacas and Quinine* (Hampshire: Macmillan, 1988), 120–34.

46. By an inadvertent set of events, some of the fourteen pounds of the seeds found their way to Madras. They were even propagated into sixty thousand seedlings, but nothing ever came of the trees; see J. H. Holland, "Ledger Bark and Red Bark," *Bulletin of Miscellaneous Information (Royal Botanic Gardens, Kew)* 1 (1932): 1–17, at 3.

47. Described in Grammacia, *Life of Charles Ledger*, 136–37.

48. Karl W. van Gorkom, *A Handbook of Cinchona Culture* (Amsterdam: J. H. Bussy, 1883), 110.

49. For a description of the many ecological challenges they faced, see Taylor, *Cinchona in Java*, 50; and Goss, *Floracrats*, 55.

50. Cinchona has what botanists technically call heterostyly, which allows it to increase cross-pollination. With so many varieties of plants on the island, this amounted to a great challenge for the Dutch, one that they solved. For a full description of the challenges and solutions, see Taylor, *Cinchona in Java*, 50.

Chapter 6

1. J. D. Hooker, "Obituary: Ward," 656.

2. Ward to Gray, 30 October 1850, Asa Gray Correspondence Files (hereafter cited as Gray Correspondence), Archives of the Gray Herbarium, Botany Libraries, Harvard University, Cambridge, MA, http://nrs.harvard.edu/urn-3:FMUS.GRA:13889116.

3. Ward to Gray, 8 October 1859, Gray Correspondence. Interestingly, this letter was sent for the first time from "The Ferns."

4. Dr. G. L. Holthouse's friendship with Ward is detailed in Thomas Lang, "On

Wardian, or Plant Cases," *Ballarat Star*, 9 April 1862, 1. A full description and sketch of Clapham appeared in Thomas Moore, "Visits to Remarkable Gardens: The Suburban Residence of N. B. Ward, Esq., at Clapham," *Gardener's Magazine*, 1851, 148–50. See also Brent Elliot, *Victorian Gardens* (London: Batsford, 1986), 32.

5. J. D. Hooker, "Obituary: Ward."

6. Ward to Gray, 20 March 1853, Gray Correspondence.

7. Ibid.

8. J. D. Hooker to Darwin, 29 March 1864, Darwin Correspondence Project, Letter no. 4439, accessed 18 August 2017, http://www.darwinproject.ac.uk/DCP-LETT-4439.

9. Ward to Gray, 13 August 1851, Gray Correspondence.

10. Ward to Gray, 11 March 1853, Gray Correspondence.

11. Described in Ward to J. D. Hooker, 4 March 1861, in Letters to J. D. Hooker, JDH1/1/2, fol. 81, RBGK.

12. Ward to Gray, 11 March 1853, Gray Correspondence.

13. Ward and Williams's involvement is recounted in Ward to W. J. Hooker, 19 December 1842, DC, vol. 18B, fol. 225, RBGK. See Joseph Paxton, "On the Culture of the *Musa cavendishii*, as Practised at Chatsworth," *Gardener's Magazine* 13 (1837): 141–42, at 142; and Joseph Paxton, "*Musa cavendishii*," *Paxton's Magazine of Botany* 3 (1837): 51–62.

14. Ward to W. J. Hooker, 9 November 1842, DC, fol. 233, RBGK.

15. Letter from Ward to the Linnaean Society read 1 November 1842, *Proceedings of the Linnean Society of London* 1 (1849): 157.

16. A. W. Murray, *Forty Years' Mission Work in Polynesia and New Guinea, from 1835 to 1875* (London: James Nisbet, 1876), 271; and Ebenezer Prout, *Memoirs of the Life of the Reverend John Williams* (London: John Snow, 1846), 149.

17. On the spread of bananas see Vaughan MacCaughey, "The Native Bananas of the Hawaiian Islands," *Plant World* 21, no. 1 (1918): 1–12; Gerard Ward, "The Banana Industry in Western Samoa," *Economic Geography* 35, no. 2 (1959): 123–37; Valérie Kagy, Maurice Wong, Henri Vandenbroucke, Christophe Jenny, Cécile Dubois, Anthony Ollivier, Céline Cardi, Pierre Mounet, et al., "Traditional Banana Diversity in Oceania: An Endangered Heritage," *PLOS One* 11, no. 3 (2016): 1–19; and Angela Kepler and Francis G. Rust, *The World of Bananas in Hawai'i: Then and Now* (Haiku, HI: Pali-O-Waipi'o Press, 2011). Plantation bananas are cloned plants; on their ecological vulnerability, see Rob Dunn, *Never Out of Season: How Having the Food We Want When We Want It Threatens Our Food Supply and Our Future* (New York: Little, Brown, 2017).

18. Ward to Gray, 6 March 1840, Gray Correspondence.

19. "Nathaniel Bagshaw Ward, Esq., Surgeon Examined," in *First Report of the Commissioners for the Inquiry into the State of Large Towns and Populous Districts*, vol. 1 (London: Clowers & Sons, 1844), 41–45, at 45. See also "The Duty on Glass," *Lancet* 1 (22 February 1845): 214–15.

20. [John Lindley], "Editorial," *Gardeners' Chronicle*, 22 February 1845, 115.

21. Ward to Gray, Autumn 1850, Gray Correspondence.

22. "The Crystal Palace," *Victoria and Albert Museum*, London, 2016, accessed 21 September 2017, http://www.vam.ac.uk/content/articles/t/the-crystal-palace/.

23. Allen, *The Victorian Fern Craze*, 43.

24. Lynn Barber, *The Heyday of Natural History, 1820–1870* (London: Doubleday, 1984), 111; David R. Hershey, "Doctor Ward's Accidental Terrarium," *American Biology Teacher* 58, no. 5 (1996): 276–81.

25. "Wardian Cases," *Illustrated London News*, 2 August 1851, 166.

26. Detailed in N. B. Ward, letter to the editor, in "Wardian Glass Cases," *Floricultural Cabinet and Florist's Magazine*, October 1851, 260.

27. He notes that he is "most busy" on a new edition in August in 1851 during the exhibition. Ward to Gray, 13 August 1851, Gray Correspondence; Ward, *On the Growth of Plants* (1852).

28. Ward to Gray, 11 August 1852, Gray Correspondence.

29. Ward to J. D. Hooker, 13 August 1866, in Letters to J. D. Hooker, JDH2/1/21, fol. 87, RBGK.

30. Ward, *On the Growth of Plants* (1852), preface.

31. George Drower, *Gardeners, Gurus and Grubs: The Stories of Garden Inventors and Innovators* (Stroud, UK: History Press, 2001), 238.

32. Asa Gray, "Growth of Plants in Glazed Cases," in *Scientific Papers of Asa Gray*, vol. 1 (Boston: Houghton, Mifflin, 1889), 59–62. The review was originally published in 1852.

33. Stephen Ward, "Obituary: Nathaniel Bagshaw Ward," *Gardeners' Chronicle*, 20 June 1868, 655–56.

34. Sue Minter, *The Apothecaries' Garden: A History of the Chelsea Physic Garden* (Stroud, UK: Sutton, 2000).

35. Ibid., 67.

36. Ward, "Obituary."

37. "Private Court," Minute Books, 5 September 1854, 438, Archives of the Worshipful Society of Apothecaries, London.

38. Ward to Gray, 29 April 1855, Gray Correspondence.

39. "Scientific Conversazione at Apothecaries' Hall," *Illustrated London News*, 28 April 1855, 405–6, at 405.

40. Ibid., 406.

41. Ward to Gray, 29 April 1855, Gray Correspondence.

42. On arranging Hooker's plants see Ward to J. D. Hooker, 15 February 1861, in Letters to J. D. Hooker, JDH1/1/2, fol. 80, RBGK. Botany quote: Ward to Gray, 1 January 1862, Gray Correspondence.

43. Ward to Gray, Spring 1863, Gray Correspondence.

44. Ward to Gray, 25 December 1866, Gray Correspondence.

45. Ibid.; emphasis in original.

46. Ward to J. D. Hooker, 30 July 1866, in Letters to J. D. Hooker, JDH2/1/21, fol. 66, RBGK.

47. "The Inventor of Wardian Cases," *Morning Post*, 13 June 1868, 3.

48. Quoted in F. Dawtrey Drewitt, *The Romance of the Apothecaries' Garden at Chelsea* (London: Chapman & Dodd, 1922), 86.

49. Ward to Gray, 20 August 1855, Gray Correspondence.

50. "The Inventor of Wardian Cases."

51. J. D. Hooker, "Obituary: Ward."

Chapter 7

1. Thomas Moore, "Editorial," *Gardeners' Chronicle*, 15 October 1870, 1372–73; "Patent Plant Case," advertisement in William Bull, *A Wholesale List of New and Beautiful Plants* (London, 1871), 170–72.

2. E. J. Wilson, *West London Nursery Gardens: The Nursery Gardens of Chelsea, Fulham, Hammersmith, Kensington and a Part of Westminster, Founded before 1900* (London: Fulham & Hammersmith Historical Society, 1982).

3. Advertisement in William Bull, *A Retail List of New and Rare Beautiful Plants* (London: William Bull, 1870).

4. "Mr William Bull's New Plant Establishment, Chelsea," *The Australasian*, 13 September 1873, 26.

5. Moore, "Editorial"; Bull, "Patent Plant Case."

6. Bull, "Patent Plant Case."

7. Stuart McCook, "Ephemeral Plantations: The Rise and Fall of Liberian Coffee, 1870–1900," in *Comparing Apples, Oranges, and Cotton: Environmental Histories of the Global Plantation*, ed. Frank Uekötter (Frankfurt: Campus, 2014), 85–112.

8. See "Death of Mr. William Bull," *Hampshire Chronicle*, 14 June 1902, 5.

9. See William Bull, *A Wholesale List of New Beautiful and Rare Plants* (London: William Bull, 1868), 2.

10. Nurseries (Botanical), Cuttings File, Kensington and Chelsea Local Studies Library, Chelsea.

11. Wilson, *West London Nursery Gardens*, 48–55.

12. Sue Shephard, *Seeds of Fortune: A Gardening Dynasty* (London: Bloomsbury, 2003); Shirley Heriz-Smith, "James Veitch & Sons of Chelsea and Robert Veitch & Son of Exeter, 1880–1969," *Garden History* 21, no. 1 (1993): 91–109.

13. E. O. Michy to William Jameson, 15 November 1843, fol. 219, DC, RBGK; William Jameson to James Veitch, 24 November 1843, fol. 159, DC, RBGK.

14. Quoted in Shephard, *Seeds of Fortune*, 98.

15. Ibid.

16. James Veitch Jr. to W. J. Hooker, 10 April 1848, fol. 586, DC, RBGK.

17. William Hooker, "Isonandra Gutta," *London Journal of Botany* 6 (1847): 464–65; James Collins, *Report on the Gutta Percha of Commerce* (London: Allen, 1878), in Miscellaneous Report Malay-Rubber, 1852–1908, MR/336, RBGK. On its destruction see Berthold Seemann, "The Taban-Tree," in Miscellaneous Report Malay-Rubber, 1852–1908, MR/336, RBGK. See also John Tully, *The Devil's Milk: A Social History of Rubber* (New York: Monthly Review Press, 2011).

18. J. G. Veitch, "Extracts from Mr. Veitch's Letters on Japan," letter dated 13 August 1860, *Gardeners' Chronicle*, 15 December 1860, 1104.

19. Ibid., 1126.

20. Robert Fortune, "Ward's Plant Cases," *Gardeners' Chronicle*, 27 June 1868, 608.

21. Fortune, *Yedo and Peking*, 147–48.

22. J. G. Veitch, "Extracts from Mr. Veitch's Letters on Japan," letter dated 20 October 1860, *Gardeners' Chronicle*, 12 January 1861, 25.

23. Shephard, *Seeds of Fortune*, 144.

24. Reproduced in James H. Veitch, *Hortus Veitchii* (London: James Veitch & Sons, 1906), 51.

25. Parsons & Co., "Japanese Trees," *Horticulturalist* 17 (1862): 186–87, at 186.

26. "New Japanese Plants," *Magazine of Horticulture and Botany* 27, no. 9 (September 1861): 412–15.

27. Stephen A. Spongberg, "The First Japanese Plants for New England," *Arnoldia* 50, no. 3 (1990): 2–11; John L. Creech, "Expeditions for New Horticultural Plants," *Arnoldia* 26, no. 8 (1966): 49–53.

28. Kristina A. Schierenbeck, "Japanese Honeysuckle (*Lonicera japonica*) as an Invasive Species: History, Ecology, and Context," *Critical Reviews in Plant Sciences* 23, no. 5 (2004): 391–400.

29. Parsons & Co., "Japanese Trees," 187.

30. James M. Howe, "George Rogers Hall, Lover of Plants," *Journal of the Arnold Arboretum* 4, no. 2 (1923): 91–98.

31. Whittingham, *Fern Fever*, 104.

32. William Robinson, "Notes on Gardens No. IV: Backhouse's Nurseries, York," *Gardeners' Chronicle*, 5 March 1864, 221.

33. Margaret Flanders Darby, "*Un*natural History: Ward's Glass Cases," *Victorian Literature and Culture* 35, no. 2 (2007): 635–47.

34. James Backhouse Jr. to James Backhouse Walker, 22 November 1860, in Letters from James Backhouse (Junior) in York, England to James Backhouse Walker in Tasmania, 1860–1871: Uncatalogued Walker Letters (hereafter cited as Backhouse Letters), University of Tasmania Library Special and Rare Materials Collection, Australia, accessed 16 July 2018, http://eprints.utas.edu.au/3238/.

35. Ibid. The details of the case are contained in this letter.

36. James Backhouse Jr. to James Backhouse Walker, 24 August 1861, Backhouse Letters.

37. See Whittingham, *Fern Fever*, 101–2.

38. David Allen, "Tastes and Crazes," in *Cultures of Natural History*, ed. N. Jardine, J. A. Secord, and E. C. Spray (Cambridge: Cambridge University Press, 1996), 400. See also Allen, *The Victorian Fern Craze*, 56–64; Whittingham, *Fern Fever*, 119.

39. Paul Fox, *Clearings: Six Colonial Gardeners and Their Landscapes* (Carlton, Victoria: Miegunyah Press, 2005), 42.

40. Thomas Lang, "On Wardian, or Plant Cases," *Ballarat Star*, 9 April 1862, 1.

41. Ibid., 1; Thomas Lang, *Catalogue of Plants Cultivated for Sale by Thomas Lang & Co* (Melbourne, 1868), iv.

42. Ibid., iv.

43. Thomas Lang, *Catalogue of Plants Cultivated for Sale by Thomas Lang & Co* (Melbourne, 1870), 5.

44. On *Nepenthes* in Wardian cases, see Veitch, *Hortus Veitchii*, 483.

45. Michelle Payne, *Marianne North: A Very Intrepid Painter* (Kew: Royal Botanic Gardens, Kew, 2016).

46. Marianne North, *Recollections of a Happy Life* (London: Macmillan, 1893), 251.

47. Quoted in "Nepenthes northiana," *Gardeners' Chronicle*, 3 December 1881, 771.

48. Shephard, *Seeds of Fortune*, 199.

49. D. Schnell, P. Catling, G. Folkerts, C. Frost, R. Gardner, et al., "*Nepenthes northiana*," The IUCN Red List of Threatened Species (2000), accessed 10 August 2017; R. B. Simpson, "Nepenthes and Conservation," *Curtis's Botanical Magazine* 12, no. 2 (1995): 111–18; Anthea Phillips and Anthony Lamb, *Pitcher-Plants of Borneo* (Kew: Royal Botanic Gardens, Kew, 1996).

50. Justin Buckley, "In the Garden," *National Trust Victoria*, February 2016, 5.

51. J. P. Bailey and A. P. Connolly, "Prize-Winners to Pariahs: A History of Japanese Knotweed *S.l.*," *Watsonia* 23 (2000): 93–110.

52. Richard Mack, "Catalog of Woes," *Natural History* 99, no. 3 (1990): 44–53, at 45.

53. The figures are taken from Andrew Liebhold, Eckehard Brockerhoff, Lynn Garrett, Jennifer Parke, and Kerry O. Britton, "Live Plant Imports: The Major Pathway for Forest Insect and Pathogen Invasions of the US," *Frontiers in Ecology and the Environment* 10, no. 3 (2012): 135–43.

54. Rebecca Epanchin-Niell and Andrew M. Liebhold, "Benefits of Invasion Prevention: Effect of Time Lags, Spread Rates, and Damage Persistence," *Ecological Economics* 116 (2015): 146–53.

Chapter 8

1. This section and the next on the inward and outward flow of plants from Kew between 1842 and 1865 is from the manuscripts Record of Plants and Seeds etc. Received by the Royal Botanic Gardens, Kew, 1805–36, 1828–47, 1848–58, and 1859–67 (hereafter cited as Plant Books, Inwards, RBGK); and the Record of Plants and Seeds etc., Sent Out by the Royal Botanic Gardens, Kew, 1828–47, 1848–59, 1860–65. Ms., Archives of the Royal Botanic Gardens, Kew (hereafter cited as Plant Books, Outwards, RBGK).

2. This is listed in the Plant Books, Outwards, 1848–59, RBGK, as No. 13, p. 408. But the returning plant list shows that W. T. March returned no. 10. No other Wardian cases were delivered to March during this period. Furthermore, no. 13 was sent out with Livingstone to Eastern Africa in February 1858, and neither any of their live plant material nor no. 13 was returned until much later. It is unknown why it was listed as no. 13 heading to Jamaica; clerical error appears the most logical explanation.

3. Plant Book, Inwards, 1859–67, RBGK, pp. 67–68.

4. Among the many books on Kew are Desmond, *Kew*; Brockway, *Science and Colonial Expansion*; and Drayton, *Nature's Government*.

5. Barber, *The Heyday of Natural History*, 112.

6. See n. 1 for details on the Plant Books.

7. Because no cases were sent or received in 1841, the analysis begins in 1842.

8. Plant Books, Inwards, 1859–67, RBGK, 237.

9. Ibid., 27.

10. J. D. Hooker to Hector, 8 November 1869, in *My Dear Hector: Letters from Joseph Dalton Hooker to James Hector, 1862–1893*, ed. John Yaldwyn and Juliet Hobbs (Wellington: Museum of New Zealand, 1998), 126.

11. J. D. Hooker to W. J. Hooker, 20 April 1843, in Joseph Hooker: Correspondence 1839–1845 from Antarctic Expedition, fols. 196–200, p. 12, RBGK.

12. Plant Books, Inwards, 1859–67, RBGK, 185.

13. Memorandum (letter), J. D. Hooker to Colonial Botanic Gardens, July 1889, Wardian Cases General File, 1/W/1, RBGK.

14. Of course, Crosby's phrase "empire of the dandelion" is apt here; see Crosby, *Ecological Imperialism*. See also James Beattie, "'The Empire of the Rhododendron': Reorienting New Zealand Garden History," in *Making a New Land: Environmental Histories of New Zealand*, ed. Tom Brooking and Eric Pawson (Dunedin: Otago University Press, 2013), 241–57; B. R. Tomlinson, "Empire of the Dandelion: Ecological Imperialism and Economic Expansion, 1860–1914," *Journal of Imperial and Commonwealth History* 26, no. 2 (1998): 84–99.

15. Warren Dean, *Brazil and the Struggle for Rubber: A Study in Environmental History* (Cambridge: Cambridge University Press, 1987).

16. Clements Markham, "The Cultivation of Caoutchouc-Yielding Trees in British India," *Journal of the Society for Arts* 24 (7 April 1876): 475–81, at 476.

17. Desmond, *Kew*, 231–35.

18. Dean, *Brazil and the Struggle for Rubber*; Joe Jackson, *The Thief at the End of the World: Rubber, Power, and the Seeds of Empire* (London: Penguin, 2009); Desmond, *Kew*.

19. William Thiselton-Dyer, "Presentation of a Piece of Plate," *Kew Bulletin* 1 (1912): 64–66, at 65.

20. Robert Cross, Report on the Investigation and Collection of Plants and Seeds of the India-Rubber Trees of Para and Ceara and Balsam of Copaiba, 20 March 1877, India Office Records, IOR/L/5/70, no 50, British Library, accessed 26 November 2019, http://www.bl.uk/manuscripts/Viewer.aspx?ref=ior!l!e!5!70_no_50_f001r.

21. Ibid., details of cases made on p. 3. See also William Bull, *A Retail List of New, Beautiful and Rare Plants* (London, 1877), 39; Dean, *Brazil and the Struggle for Rubber*, 27.

22. Quoted in Dean, *Brazil and the Struggle for Rubber*, 29; see also Desmond, *Kew*, 234.

23. "Presumably . . . ," in Stanley Arden, *Report on* Hevea brasiliensis *in the Malay Peninsula* (Taiping: Perak Government Printing Office, 1902), 1, in Malay Rubber, 1852–1908, Miscellaneous Report 336, p. 124, RBGK. H. N. Ridley to William Thiselton-Dyer, 20 October 1897, in Miscellaneous Report Malay-Rubber, 1852–1908, MR/336, p. 252, RBGK.

24. Gilbert James to Finlay Muir & Co., 27 July 1906, Kew MR/760 Brazil, Balata Gum and Rubber, 1877–1908, pp. 272–73, RBGK. See also Dean, *Brazil and the Struggle for Rubber*, 53–66.

25. Compiled from Plant Books, Outwards, 1881–1895; and Plant Books, Outwards, 1896–1923.

Chapter 9

1. Drayton, *Nature's Government*, 257. Londa Schiebinger and Claudia Swan, "Introduction," in *Colonial Botany: Science, Commerce, and Politics in the Early Modern World* (Philadelphia: University of Pennsylvania Press, 2005), 1–16; Brockway, *Science and Colonial Expansion*.

2. Daniel Headrick, *The Tentacles of Progress: Technology Transfer in the Age of Imperialism, 1850–1940* (New York: Oxford University Press, 1988); Daniel Headrick, "Botany, Chemistry, and Tropical Development," *Journal of World History* 7, no. 1 (1996): 1–20; Heather Streets-Salter and Trevor Getz, *Empires and Colonies in the Modern World: A Global Perspective* (Oxford: Oxford University Press, 2016), 308–9; Robert Kubicek, "British Expansion, Empire and Technological Change," in *Oxford History of the British Empire*, vol. 3, ed. Andrew Porter and Wm. Roger Louis (London: Oxford University Press, 2001), 247–69.

3. German Embassy, London, to J. D. Hooker, 12 September 1879, in Miscellaneous Report (hereafter cited as MR): 53: Germany, RBGK, p. 2; and Walter Lack, "Kew: Ein Vorbild für Berlin-Dahlem?," in *Preußische Gärten in Europa: 300 Jahre Gartengeschichte* (Leipzig: Edition, 2007), 182–85.

4. "Create an . . . ," in Otto Hirschfeld to William Thiselton-Dyer, 9 June 1888, in MR/53: Germany, RBGK, p. 17.

5. "Interior arrangements . . . ," in ibid.; Georg Schweinfurth to William Thiselton-Dyer, 17 November 1888, in MR/53: Germany, RBGK, p. 20; William Thiselton-Dyer to Secretary of the Office of Works, 14 February 1889, in MR/53: Germany, RBGK, p. 26.

6. "The Germans . . . ," in T. V. Lister to William Thiselton-Dyer, 17 December 1888, in MR/53: Germany, RBGK, p. 19.

7. Isabelle Lévêque, Dominique Pinon, and Michel Griffon, *Le jardin d'agronomie tropicale: De l'agriculture coloniale au développement curable* (Paris: CIRAD, 2005).

8. W. D. Brackenridge, "A Historical and Descriptive Account of the Botanic Garden at Berlin," *Gardener's Magazine* 12 (June 1836): 295–310; Georg Kohlmaier, *Das Glashaus: Ein Bautypus des 19. Jahrhunderts* (Munich: Prestel, 1981).

9. Ulrike Lindner, "Trans-Imperial Orientation and Knowledge Transfers: German Colonialism in the International Context," in *German Colonialism: Fragments Past and Present* (Berlin: Deutsches Historisches Museum, 2016), 29; Hanan Sabea, "Pioneers of Empire? The Making of Sisal Plantations in German East Africa, 1890–1917," in *German Colonialism Revisited: African, Asian, and Oceanic Experiences* (Ann Arbor: University of Michigan Press, 2014), 114–29. See also Bradley Naranch, "Introduction: German Colonialism Made Simple," in *German Colonialism in a Global Age*, ed. Bradley Naranch and Geoff Eley (Durham, NC: Duke University Press, 2014), 1–18; Streets-Salter and

Getz, *Empires and Colonies*. On a specific colony, see Peter Sack, "German New Guinea: A Reluctant Plantation Colony?" *Journal de la Société des océanistes* 42, nos. 82–83 (1986): 109–27.

10. This can be observed from the many images in Ferdinand Wohltmann, *120 Kultur- und Vegetations-Bilder aus unseren deutschen Kolonien* (Berlin: Wilhelm Süsserott, 1904).

11. A. Engler, *Der Königliche Botanischen Garten und das Botanische Museum zu Berlin im Etatsjahr 1891–92* (Berlin: Julius Becker, 1892), 1. Quote from A. Engler, "Gutachten über den königlichen botanischen Garten zu Berlin und über die Frage nach seiner Verlegung," *Notizblatt des königlichen botanischen Gartens und Museums zu Berlin* 10 (1897): 301, quoted and translated in Katja Kaiser, "Exploration and Exploitation: German Colonial Botany at the Botanic Garden and Botanical Museum Berlin," in *Sites of Imperial Memory: Commemorating Colonial Rule in the Nineteenth and Twentieth Centuries* (Manchester: Manchester University Press, 2015), 225–42, at 228. See also Bernhard Zepernick, "Die Botanische Zentralstelle für die deutschen Kolonien," in *Kolonialmetropole Berlin: Eine Spurensuche*, ed. Ulrich van der Heyden and Joachim Zeller (Berlin: Berlin Edition, 2002), 107–11.

12. For 1880s figures see A. W. Eichler, *Jahrbuch des Königlichen Botanischen Gartens und des Botanischen Museums zu Berlin* (Berlin, 1881), viii–x; A. W. Eichler, *Jahrbuch des Königlichen Botanischen Gartens und des Botanischen Museums zu Berlin* (Berlin, 1884), viii–xii. Note that no yearly reports were issued between 1884 and 1889. For 1890, see Engler, *Der Königliche Botanische Garten*, 6–7. For seeds sent in 1891 (242.33 kg), see A. Engler, *Botanische Jahrbücher für Systematik, Pflanzengeschichte und Pflanzengeographie* 35 (1892): 11. On comparative figures with 1894, see A. Engler, *Jahrbuch des Königlichen Botanischen Gartens und des Botanischen Museums zu Berlin 1893–94* (Berlin: Julius Becker, 1894); A. Engler, *Jahrbuch des Königlichen Botanischen Gartens und des Botanischen Museums zu Berlin im Etatsjahr 1895–96* (Berlin: Julius Becker, 1896), 6–7.

13. "Der Botanische Garten zu Berlin," *Kolonie und Heimat* 35 (21 May 1911): 2–3, at 3, my translation.

14. G. Volkens, "Kulturerfolge des Versuchsgartens von Victoria in Kamerun mit den von der Botanischen Centralstelle in Berlin gelieferten Nutzpflanzen, nach Berichten des Direktors Dr. Preuss," *Notizblatt des königlichen botanischen Gartens und Museums zu Berlin* 2, no. 14 (1898): 159–73. The sourcing of other plants is described in a letter from Paul Preiss quoted in Gustav Meinecke, *Koloniales Jahrbuch: Beiträge und Mitteilungen aus dem Gebiete der Kolonialwissenschaft und Kolonialpraxis* (Berlin: Carl Heymanns, 1895), 81. On gutta percha, see O. Warburg, "Guttaperchakultur in Kamerun," *Tropenpflanzer* 6 (1902): 561–64.

15. A. Engler, *Bericht über den Botanischen Garten und das Botanische Museum zu Berlin im Rechnungsjahr 1901* (Halle: Waisenhauses, 1902), 5, my translation. The list of cases is compiled from A. Engler, "Bericht über die Thätigkeit der botanischen Centralstelle für die Kolonieen im Jahre 1901," *Notizblatt des königlichen botanischen Gartens und Museums zu Berlin* 3, no. 28 (24 February 1902): 176–81.

16. The figures for 1902 are from A. Engler, "Bericht über die Tätigkeit der

Botanischen Zentralstelle für die deutsche Kolonien am Königl. botanischen Garten und Museum zu Berlin im Jahre 1902," *Notizblatt des königlichen botanischen Gartens und Museums zu Berlin* 3, no. 30 (15 March 1903): 215–24. See also A. Engler, *Bericht über den Botanischen Garten und das Botanische Museum zu Berlin im Rechnungsjahr 1904* (Halle: Waisenhauses, 1905), 7; A. Engler, *Bericht über den Botanischen Garten und das Botanische Museum zu Berlin im Rechnungsjahr 1905* (Halle: Waisenhauses, 1906), 7; A. Engler, *Bericht über den Botanischen Garten und das Botanische Museum zu Berlin im Rechnungsjahr 1908* (Halle: Waisenhauses, 1909), 7; A. Engler, *Bericht über den Botanischen Garten und das Botanische Museum zu Berlin im Rechnungsjahr 1910* (Halle: Waisenhauses, 1911), 7. On Ernst Ule's expedition see "Ule's Expedition nach den Kautschuk-Gebieten des Amazonenstromes: Vierter Bericht über den Verlauf der Kautschuk-Expedition vom November 1901 bis zum März 1902," *Notizblatt des königlichen botanischen Gartens und Museums zu Berlin* 4, no. 32 (30 August 1903): 92–98. The figures are from Otto Lutz, "Botanische Zentralstelle für die Kolonien," *Kolonie und Heimat* 22 (1910): 2–3. Curators at the Botanic Garden and Botanical Museum Berlin give the estimate between 1891 and 1907 as 16,500 plants moved in Wardian cases; see Zepernick, "Die Botanische Zentralstelle," 109–10; and Kathrin Grotz and H. Walter Lack, "Wardsche Kästen," *Museums Journal* 4 (2010), accessed 27 November 2019, https://www.museumsportal-berlin.de/de/magazin/blickfange/wardsche -kasten-ein-dachbodenfund/. See also Katja Kaiser, "Wardian Case for Shipping Live Plants, Berlin, around 1900," in *German Colonialism: Fragments Past and Present* (Berlin: Deutsches Historisches Museum, 2016), 202.

17. Quoted in Nancy Y. W. Tom, *The Chinese in Western Samoa, 1875–1985: The Dragon Came from Afar* (Apia: Western Samoa Historical Trust, 1986), 1; Malama Meleisea and Penelope Schoeffel, "Before and After Colonisation: Germany in Samoa," in *German Colonialism: Fragments Past and Present* (Berlin: Deutsches Historisches Museum, 2016), 118–27.

18. Tom, *Chinese in Western Samoa,* 1–11. For a similar treatment of workers in Africa, see the discussions in W. A. Crabtree, "German Colonies in Africa," *Journal of the Royal African Society* 14, no. 53 (1914): 1–14.

19. "German Colonies in Tropical Africa," *Bulletin of Miscellaneous Information (Royal Botanic Gardens, Kew)* 96 (1894): 410–12, at 412. Stuart McCook, "Global Rust Belt: *Hemileia vastatrix* and the Ecological Integration of World Coffee Production since 1850," *Journal of Global History* 1, no. 2 (2006): 177–95.

20. "Die Kolonien," *Samoanische Zeitung,* 15 August 1914, 3.

21. The key work here is Christophe Bonneuil and Mina Kleiche, *Du jardin d'essais colonial à la station expérimentale, 1880–1930: Éléments pour une histoire du CIRAD* (Paris: CIRAD, 1993).

22. Jean Dybowski, *Le Congo meconnu* (Paris: Hachette, 1912).

23. "Le jardin colonial," *La dépêche coloniale,* 15 August 1903, 199, my translation. The journalist was following the lead of Dybowski in making these comments.

24. Lévêque, Pinon, and Griffon, *Le jardin d'agronomie tropicale.*

25. Bernard Verlot, "Le botaniste herborisant," *La Nature* (1878): 283–87. The

Vilmorin-Andrieux advertisement appears regularly in the French colonial journal; see, for example, the back pages of *La dépêche coloniale*, 15 August 1903. Gutta percha is discussed in J. Paul Trouillet, "Le Congo Français," *La dépêche coloniale*, 30 April 1902, 11.

26. For the British interest in Madagascar, see J. Gilbert Baker, "Lecture on Madagascar," *Thames Valley Times*, 11 January 1888.

27. P. Danthu, H. Razakamanarivo, L. Razafy Fara, P. Montagne, B. Deville-Danthu, and E. Penot, "When Madagascar Produced Natural Rubber: A Brief, Forgotten yet Informative History," *Bois et forêts des tropiques* 328, no. 2 (2013): 27–43. The botanical products of Madagascar are described in "Madagascar," MR/558, RBGK.

28. Joseph Galliéni, *Rapport d'ensemble du Général Galliéni sur la situation générale de Madagascar*, vol. 2 (Paris: Imprimerie des Journaux Officiels, 1899), 42–43.

29. "L'agriculture á Madagascar," *La dépêche coloniale*, 31 October 1903, 266–80.

30. Robert Aldrich, *Greater France: A History of French Overseas Expansion* (Hampshire: Macmillan, 1996).

31. Robert Aldrich, *Vestiges of Colonial Empire in France* (London: Palgrave, 2004), 61–67, at 65; Lévêque, Pinon, and Griffon, *Le jardin d'agronomie tropicale*, 79–87.

32. Louis Vernet, "Une visite au jardin colonial," *La dépêche coloniale*, 30 June 1909, 155–66. The quote on 158 is translated in Aldrich, *Vestiges*, 66.

33. "La plupart des plantes économiques rassemblées à l'Ivoloina ont été envoyées en serres portatives par le Jardin Colonial" (Vernet, "Une visite," 164).

34. "Exposition de Bruxelles: Section Coloniale Française," *La dépêche coloniale*, 15 April 1910, 7. The estimate is based on one hundred plants in each case. This is the most conservative number; Vernet estimated between sixty and a hundred plants in each case. See Vernet, "Une visite," 163.

35. Christophe Bonneuil, "'Mise en valeur' de l'empire colonial et naissance de l'agronomie tropicale," in Bonneuil and Mina, *Du jardin d'essais colonial à la station expérimentale*, 7.

Chapter 10

1. John McNeil has discussed the rise of environmental management in the American imperial context; see J. R. McNeil, "Introduction: Environmental and Economic Management," in *The Colonial Crucible: Empire in the Making of the Modern American State*, ed. Alfred W. McCoy and Francisco A. Scarano (Madison: University of Wisconsin Press, 2009), 475–78, at 476.

2. "IPCC 65th Anniversary," International Plant Protection Convention, accessed 21 November 2019, https://www.ippc.int/en/themes/ipp-65th-anniversary/.

3. Alan MacLeod, Marco Pautasso, Mike J. Jeger, and Roy Haines-Young, "Evolution of the International Regulation of Plant Pests and Challenges for Future Plant Health," *Food Security* 2, no. 1 (2010): 49–70; and Liebhold and Griffin, "The Legacy of Charles Marlatt."

4. These include the blueberry, cranberry, pecan, squash, sugar maple, sunflower,

and tobacco, see Peter Coates, *American Perceptions of Immigrant and Invasive Species: Strangers on the Land* (Berkeley: University of California Press, 2006), 81.

5. The purpose of the division is described in Beverley Galloway, "Immigrant Plants Hold Large Place among U.S. Crops," in *Yearbook of Agriculture 1928* (Washington, DC: Government Printing Office, 1929), 379.

6. See Amanda Harris, *Fruits of Eden: David Fairchild and America's Plant Hunters* (Gainesville: University of Florida Press, 2015), 243.

7. David Fairchild, *Systematic Plant Introduction* (Washington, DC: USDA, 1898), 7–8.

8. On early introductions to the United States, see Howard Hyland, "History of U.S. Plant Introduction," *Environmental Review* 2, no. 4 (1977): 26–33.

9. Bartlett, "Reports of the Wilkes Expedition," 677.

10. Much of the following section discusses David Fairchild, but Russia also had an important plant hunter at this time who was equally interested in plant breeding: Nicolay Vavilov. See Nicolay I. Vavilov, *Five Continents*, trans. Doris Löve (Rome: IPGRI, 1997).

11. US Department of Agriculture, "Seeds and Plants Imported: Inventory from December 1903 to December 1905" (hereafter all titles cited as USDA, "Plant Inventory"), 11 (1907): 5. For all plant accession records see https://naldc-legacy.nal.usda .gov/naldc/collections.xhtml. Like most plant lists kept by botanical institutions, the USDA did not list how plants arrived; therefore, it is unknown how many Wardian cases were used. An example is USDA, "Plant Inventory: 1 July to 30 September 1908," 16 (1909), where Fairchild discusses ("Introduction," p. 5) how a Wardian case of lychees arrived at the Office, however, in the inventory's description, no. 23364–23366 (p. 10), no Wardian case is listed, only the types of plants introduced. The 1928 accessions are listed in USDA, "Plant Inventory: October 1 to December 31, 1928," 97 (1930).

12. Galloway, "Immigrant Plants," 379.

13. Walter F. Burton, "Treasure Hunters Comb Earth for Priceless Plants," *Popular Science Monthly*, August 1932, 35.

14. Quoted in ibid., 112.

15. Fairchild, introduction to USDA, "Plant Inventory: October 1 to December 31, 1908," 17 (1909): 7.

16. Burton, "Treasure Hunters," 34.

17. See Stuart McCook, "'The World Was My Garden': Tropical Botany and Cosmopolitanism in American Science, 1898–1935," in McCoy and Scarano, *Colonial Crucible*, 499–507; and Paul S. Sutter, "Tropical Conquest and the Rise of the Environmental Management State: The Case of the U.S. Sanitary Efforts in Panama," in McCoy and Scarano, *Colonial Crucible*, 317–26.

18. David Fairchild, "Two Expeditions after Living Plants," *Scientific Monthly* 26, no. 2 (1928): 97–127, at 112.

19. Essential oils in David Fairchild, *Bulletin of Foreign Plant Introductions* 12 (15–28 February 1909): 3; lychees in Fairchild, "Plant Inventory," *Bulletin of Foreign Plant Introductions* 16 (1909): 5; garcinias in Fairchild, *Bulletin of Foreign Plant Introductions*

18 (1909): 8; Buitenzorg plants in Fairchild, *Bulletin of Foreign Plant Introductions* 26 (1910): 4 and 48 (1910): 5; Buitenzorg mangoes in Fairchild, *Bulletin of Foreign Plant Introductions* 86 (1913): 665; mangroves in Fairchild, "Plant Inventory," *Bulletin of Foreign Plant Introductions* 36 (1913): 62; on Mrs. Arthur Curtis James, in Fairchild, "Plant Inventory," *Bulletin of Foreign Plant Introductions* 196 (1922): 1785; Ceylon quote in Wilson Popenoe to David Fairchild, 1 October 1912, in David Fairchild, *Bulletin of Foreign Plant Introductions* 81 (1912): 612–13.

20. Wilson Popenoe to David Fairchild, 1 October 1912, in David Fairchild, *Bulletin of Foreign Plant Introductions* 81 (1912): 612–13. The Hartless quotes in the next paragraph appear in the same edition of the *Bulletin*; see A. C. Hartless to David Fairchild, 9 October 1912, 615–16.

21. Fairchild to W. A. Taylor, 25 July 1916, in "Typescript of South China Explorations," Frank N. Meyer Collection, Collection no. 295, pp. 1–8, USDA, National Agricultural Library, Beltsville, MD, accessed 23 November 2019, https://www.nal .usda.gov/exhibits/speccoll/exhibits/show/frank-meyer.

22. Hyland, "History," 30; and G. A. Weber, *The Plant Quarantine and Control Administration: Its History, Activities and Organization* (Washington, DC: Brookings Institute, 1930).

23. Marlatt quotes in Marlatt, "Losses Caused by Imported Tree and Plant Pests," *American Forestry* 23 (1917): 75–80, at 76. Marlatt is referring to W. Dwight Pearce, *A Manual of Dangerous Insects Likely to be Introduced in the United States through Importations* (Washington, DC: Government Printing Office, 1918).

24. Pierce, *Manual*, 58.

25. Marlatt, "Losses," 79.

26. David Fairchild, "The Independence of American Nurseries," *American Forestry* 23 (1917): 213–16.

27. Ibid., 216.

28. Pauly, *Fruits and Plains*, 151–53.

29. Coates, *American Perceptions*, 101–7; and Pauly, *Fruits and Plains*, 154–58.

30. Kent Beattie, "The Operation of Quarantine no. 37," *Journal of Economic Entomology* 14, no. 2 (1921): 201–5.

31. Marcia Kreith and Deborah Golino, "Regulatory Framework and Institutional Players," in *Exotic Pests and Diseases: Biology and Economics for Biosecurity*, ed. Daniel A. Sumner (Ames: Iowa State Press, 2003), 20.

32. Charles Marlatt, "Protecting the United States from Plant Pests," *National Geographic*, August 1921, 205–18, at 210 and 214.

33. In 1909, a decade before the introduction of Quarantine no. 37, Marlatt and Fairchild came into conflict over the Japanese cherry tree incident; see Phillip Pauly, "The Beauty and Menace of the Japanese Cherry Trees: Conflicting Visions of American Ecological Independence," *Isis* 87, no. 1 (1996): 51–74. In this classic essay Pauly describes the incident, which placed American entomologists and plant explorers at odds with each other. He traces how these debates over flora and fauna, and the strin-

gent absolutist and exclusionist policies promoted by Marlatt, had a major influence on human immigration policies at the time and later.

34. Russell Woglum, *A Report of a Trip to India and the Orient in Search of the Natural Enemies of the Citrus White Fly*, Bureau of Entomology, Bulletin no. 120 (Washington, DC: USDA, 1913), 35.

35. Stephen Hamblin, "Plants and Policies," *Atlantic Monthly*, March 1925, 353–62, at 350.

36. *An Appeal to Every Friend of American Horticulture*, report from the Conference of the Representatives of Horticultural and Other Societies, held at the American Museum of Natural History, 15 June 1920, in Letters of J. Horace MacFarland to Charles Sargent, Archives of the Arnold Arboretum, Harvard University.

37. "Prohibition of Imports of Plants into the United States," *Gardeners' Chronicle*, 15 February 1919, 76.

38. Minutes of the Royal Horticultural Society, 14 January 1919, 11 February, 11 March, Minutes Book, 182, 191, and 204–5; quote at 191. Archives of the Royal Horticultural Society, Lindley Library, London.

39. Weber, *Plant Quarantine*.

40. Beverly T. Galloway, *How to Collect, Label, and Pack Living Plant Material for Long-Distance Shipment*, Department Circular 323 (Washington, DC: USDA, 1924), at 1–2; on the Wardian case, see 10. See also P. D. Peterson and C. L. Campbell, "Beverly T. Galloway: Visionary Administrator," *Annual Review of Phytopathology* 35 (1997): 29–43.

41. Fairchild, "Two Expeditions," 125.

42. "Plant Research Yacht Lands Large Caribbean Collection," *California Garden* 24, no. 5 (1932): 9.

43. T. Ralph Robinson, "Safeguarding the Introduction of Citrus Plants through Improved Quarantine Methods," *Proceedings of the Florida State Horticultural Society* 36 (1923): 26–32, at 28; and Walter Swingle, T. Ralph Robinson, and Eugene May, *Quarantine Procedure to Safeguard the Introduction of Citrus Plants: A System of Aseptic Plant Propagation* (Washington, DC: USDA, 1924), 15.

Chapter 11

1. Stuart L. Pimm, "What Is Biodiversity?," in *Sustaining Life: How Human Health Depends on Biodiversity*, ed. Eric Chivan and Aaron Bernstein (New York: Oxford University Press, 2008), 6–12. The inside of Kew's case (just the box that held the soil) in the Economic Botany collection was measured by the author as 70 × 25 × 83 cm = 14,250 cm^3, equal to about 0.15 m^3 or 5 ft^3. Each box was filled to at least halfway up the surface area, making the volume of soil about 2.5 ft^3. This is a very conservative estimate.

2. E. O. Wilson, foreword, in David Littschwager, *A World in One Cubic Foot: Portraits in Biodiversity* (Chicago: University of Chicago Press, 2012).

3. Ken Killham, *Soil Ecology* (Cambridge: Cambridge University Press, 1994).

4. For a full description of the worm, see H. N. Moseley, "Description of a New

Species of Land-Planarian from the Hothouses at Kew Gardens," *Journal of Natural History* 1, no. 3 (1878): 237–39.

5. Ibid., 238.

6. Leigh Winsor, Peter Johns, and Gary Barker, "Terrestrial Planarians (Platyhelminthes: Tricladida: Terricola) Predaceous on Terrestrial Gastropods," in *Natural Enemies of Terrestrial Molluscs*, ed. Gary Barker (Wallingford: CABI, 2004), 227–78.

7. Ronald Sluys, "Invasion of the Flatworms," *American Scientist* 104, no. 5 (2016): 288–95.

8. Winsor, Johns, and Barker, "Terrestrial Planarians," 241.

9. Leigh Winsor, "A Revision of the Cosmopolitan Land Planarian *Bipalium kewense*, Moseley, 1878 (Turbellaria: Tricladida: Terricola)," *Zoological Journal of the Linnean Society* 79, no. 1 (1983): 61–100, countries listed at 80.

10. Winsor, Johns, and Barker, "Terrestrial Planarians," 241.

11. Jean-Lou Justine, Leigh Winsor, Delphine Gey, Pierre Gros, and Jessica Thévenot, "The Invasive New Guinea Flatworm *Platydemus manokwari* in France, the First Record for Europe: Time for Action Is Now," *PeerJ* 2 (4 March 2014): e297, p. 3.

12. Ibid., 9.

13. Jean-Lou Justine, Leigh Winsor, Patrick Barrière, Crispus Fanai, Delphine Gey, Andrew Kien Han Wee, Giomara La Quay-Velázquez, Benjamin P. Y.-H. Lee, et al., "The Invasive Land Planarian *Platydemus manokwari* (Platyhelminthes, Geoplanidae): Records from Six New Localities, Including the First in the USA," *PeerJ* 3 (2015): e1037, 1–20.

14. Worms weren't the only organisms carried in the soil. One New Zealand botanist was keenly aware by the late nineteenth century that many of the soil heaps spread around the country held "weeds [that] came from Wardian Cases sent out here"; see William Travers, quoted in W. M. Maskell, "Abstract of Annual Report," *Transactions and Proceedings of the Royal Society of New Zealand* 28 (1895): 745–46, at 745.

15. Specifically, diseases and pathogens are classified as either abiotic disorders, prokaryote diseases, viral diseases, fungal diseases, oomycete diseases, fungal-like organisms, nematode diseases, or parasitic plants. For more detailed descriptions, see the American Phytopathological Society, *Diseases and Pathogens*, accessed June 20, 2019, https://www.apsnet.org/edcenter/disandpath/Pages/default.aspx.

16. This discussion is taken from McCook, "Global Rust Belt." See also Dunn, *Never Out of Season*, 6–7.

17. First reported in May 1868. *Hemileia vastatrix* first described in "Editorial," *Gardeners' Chronicle*, 6 November 1869, 1157. See also Maria do Céu Silva, Victor Várzea, Leonor Guerra-Guimarães, Helena Gil Azinheira, Diana Fernandez, Anne-Sophie Petitot, Benoit Bertrand, Philippe Lashermes, and Michel Nicole, "Coffee Resistance to the Main Diseases: Leaf Rust and Coffee Berry Disease," *Brazilian Journal of Plant Physiology* 18, no. 1 (2006): 119–47.

18. Paul R. Miller, review of John Stevenson, *Fungi of Puerto Rico and the American Virgin Islands*, *Taxon* 24, no. 4 (1975): 522.

19. Dunn, *Never Out of Season*.

20. P. M. Austin Bourke, "Emergence of Potato Blight, 1843–46," *Nature* 203, no. 4947 (1964): 805; and S. B. Goodwin, B. A. Cohen, and W. E. Fry, "Panglobal Distribution of a Single Clonal Lineage of the Irish Potato Famine Fungus," *Proceedings of the National Academy of Sciences of the United States of America* 91, no. 24 (1994): 11591.

21. S. A. M. H. Naqvi, "Diagnosis and Management of Certain Important Fungal Diseases of Citrus," in *Diseases of Fruits and Vegetables: Diagnosis and Management,* ed. S. A. M. H. Naqvi (Dordrecht: Kluwer Academic, 2004), 247–90, at 249.

22. H. S. Fawcett, *Citrus Diseases and Their Control* (New York: McGraw-Hill, 1926), 146–47.

23. Ibid., 147.

24. C. M. Brasier, "The Biosecurity Threat to the UK and Global Environment from International Trade in Plants," *Plant Pathology* 57, no. 5 (2008): 792–808.

25. P. Hennings. "Über die auf *Hevea*-Arten bisher beobachteten parasitischen Pilze," *Notizblatt des königlichen botanischen Gartens und Museums zu Berlin* 4, no. 34 (1904): 133–38.

26. Dean, *Brazil and the Struggle for Rubber*; and McCook, "'Squares of Tropic Summer,'" 214.

27. Susan Freinkel suggests that that Parsons's shipments of 1876 were most likely the earliest source of chestnut blight, though this is speculative; see Freinkel, *American Chestnut: The Life, Death, and Rebirth of a Perfect Tree* (Berkeley: University of California Press, 2007), 68. Hall's shipment brings the arrival of Japanese chestnuts back by a decade, although whether or not these were infected is unknown.

28. Albert Koebele, "Report of a Trip to Australia Made under Direction of the Entomologist to Investigate the Natural Enemies of the Fluted Scale," *U.S. Department of Agriculture, Division of Entomology Bulletin* 21 (1890): 1–32.

29. "The Fluted Scale-Insect (Icerya Purchasi, Maskell.)," *Bulletin of Miscellaneous Information (Royal Botanic Gardens, Kew)* 32 (1889): 191–216.

30. Ibid., 192.

31. Known at the time as *Verdalia cardinalis*.

32. Koebele, "Report of a Trip to Australia," 13.

33. Koebele to C. V. Riley, undated (first letter from Australia), in C. V. Riley, *Report of the Entomologist for the Year 1886* (Washington, DC: Government Printer, 1886–88), 90–91.

34. Paul DeBach and David Rosen, *Biological Control by Natural Enemies* (Cambridge: Cambridge University Press, 1991), 140–49; and Susanna Iranzo, Alan L. Olmstead, and Paul W. Rhode, "Historical Perspectives on Exotic Pests and Diseases in California," in *Exotic Pests and Diseases: Biology and Economics for Biosecurity,* ed. Daniel A. Sumner (Ames: Iowa State Press, 2008), 55–67.

35. DeBach and Rosen, *Biological Control by Natural Enemies,* 140.

36. Woglum also had ladybugs (*Cryptognatha flavescens*) in the Wardian cases. See M. Rose and P. DeBach, "Citrus Whitefly Parasites Established in California," *California Agriculture* (July–August 1981): 21–23.

37. L. O. Howard to James Wilson, 11 September 1912, in Woglum, *Report of a Trip*

to India and the Orient, front matter. The suggestions of both Fairchild and Marlatt appear on p. 35.

38. Curtis Clausen and Paul A. Berry, *Citrus Blackfly in Asia, and the Importation of Its Natural Enemies into Tropical America,* Technical Bulletin 320 (Washington, DC: USDA, 1932).

39. Ibid., 52.

40. Jodi Frawley, "Prickly Pear Land: Transnational Networks in Settler Australia," *Australian Historical Studies* 38, no. 130 (2007): 323–38; and Jeffrey C. Kaufmann, "Prickly Pear Cactus and Pastoralism in Southwest Madagascar," *Ethnology* 43, no. 4 (2004): 345–61.

41. Peter MacOwan, "Prickly Pear in South Africa," *Kew Bulletin* 19 (1888): 165–73, at 165.

42. Leonie Seabrook and Clive McAlpine, "Prickly Pear," *Queensland Historical Atlas* 1 (2010), accessed 27 November 2019, http://www.qhatlas.com.au/content/prickly-pear.

43. T. Harvey Johnston and Henry Tryon, *Report of the Queensland Prickly-Pear Travelling Commission* (Brisbane: Government Printer, 1914). As early as 1914 promising parasitic insects were found, including two cochineal insects and the cactoblast moth, but breeding efforts were put on hold during the war.

44. Jodi Frawley, "A Lucky Break: Contingency in the Storied Worlds of Prickly Pear," *Continuum: Journal of Media and Cultural Studies* 28, no. 6 (2014): 760–73, at 764.

45. Alan Dodd, *The Progress of Biological Control of Prickly-Pear in Australia* (Brisbane: Government Printer, 1929).

46. Alan Dodd, *The Biological Campaign against Prickly-Pear* (Brisbane: Government Printer, 1940), 72.

47. Dodd, *Progress,* 17. Much of the following discussion is taken from 17–18 and from Dodd, *Biological Campaign,* 72–73.

48. Dodd, *Biological Campaign,* 72.

49. DeBach and Rosen, *Biological Control by Natural Enemies,* 167–68.

50. Ibid., 133–35; J. D. Tothill, T. H. C. Taylor, and R. W. Paine, *The Coconut Moth in Fiji: A History of Its Control by Means of Parasite* (London: Imperial Bureau of Entomology, 1930); Armand Kuris, "Did Biological Control Cause Extinction of the Coconut Moth, *Levuana iridescens,* in Fiji?," *Biological Invasions* 5, no. 1 (2003): 133–41; and Daniel Simberloff and Peter Stiling, "Risks of Species Introduced for Biological Control," *Invasion Biology* 78, no. 1 (1996): 185–92.

Conclusion

1. William Andrew Archer, *Collecting Data and Specimens for Study of Economic Plants,* Misc. Publication No. 568 (Washington, DC: USDA, 1945), 41.

2. On the popularity of hardy plants, see Charles Sargent to Harry Veitch, 5 February 1908, in Charles Sargent Letter Books, vol. 3, p. 652, Arnold Arboretum Archives, Harvard University. The other, more practical matter that effected this shift was the

unavailability of fuel to heat greenhouses, which was most noticeable during and after the First World War.

3. J. Lemaistre to Edward Salisbury, 4 November 1948, Wardian Case General Files 1/W/1, RBGK. See also Thomas Garner James, "Kew: The Commoners' Royal Garden," *National Geographic*, April 1950, 479–506; and Desmond, *Kew*, 276 and 354. The 1987 *Plants under Glass* exhibition is detailed in Christine Brandt to Peter Wood, 5 February 1987, Wardian Case General Files 1/W/1, RBGK.

4. "The History of Botanic Gardens," *Botanic Gardens Conservation International*, accessed 15 December 2017, https://www.bgci.org/resources/history/; and Roy Ballantyne, Jan Packer, and Karen Hughes, "Environmental Awareness, Interests and Motives of Botanic Gardens Visitors: Implications for Interpretive Practice," *Tourism Management* 29, no. 3 (2008): 439–44.

5. Geoffrey C. Marks and Ian W. Smith, *The Cinnamon Fungus in Victorian Forests: History Distribution Management and Control* (Melbourne: Department of Conservation and Environment, 1991); Wei Y. Hee, Pernelyn S. Torenna, Leila M. Blackman, and Adrienne R. Hardham, "*Phytophthora cinnamomi* in Australia," in *Phytophthora: A Global Perspective*, ed. K. Lamour (Cambridge, MA: CABI, 2013), 124–34; and Gretna Weste and G. C. Marks, "The Distribution of *Phytophthora cinnamomi* in Victoria," *Transactions of the British Mycological Society* 63, no. 3 (1974): 559–72.

6. J. A. Crooks and M. E. Soulé, "Lag Times in Population Explosions of Invasive Species: Causes and Implications," in *Invasive Species and Biodiversity Management*, ed. Odd Terje Sandlund, Peter Johan Schei, and Aslaug Viken (Dordrecht: Kluwer Academic, 1999), 103–25; and Epanchin-Niell and Liebhold, "Benefits of Invasion Prevention."

7. Mack and Lonsdale, "Humans as Global Plant Dispersers." See also Reuben P. Keller, Juergen Geist, Jonathan M. Jeschke, and Ingolf Kuhn, "Invasive Species in Europe: Ecology, Status, and Policy," *Environmental Sciences Europe* 23 (2011): 1–17; and David Pimentel, Rodolfo Zuniga, and Doug Morrison, "Update on the Environmental and Economic Costs Associated with Alien-Invasive Species in the United States," *Ecological Economics* 52, no. 3 (2005): 273–88.

Index

....

Page numbers in italics refer to figures.